HOW TO BUILD WITH STEEL

by ROBERT PARDUN

SHIRE PRESS

LIBRARY OF CONGRESS # 99-096622
ISBN: 0-918828-19-8

SHIRE PRESS
26873 Hester Creek Road
Los Gatos, CA 95033
(408)353-4253

CONTENTS

INTRODUCTION

As is the case with most steel fabricators, I started out as a welder. I observed that the blueprints for the objects to be built were given to fabricators who then gathered the required material, cut it to length, prepared each piece by drilling, cutting, adding stiffeners and the like, tack welded the pieces together into the final structure, and gave it to a welder who did the final welding. After the piece was welded the fabricator would remove welding distortion, have final machining done if required and install the machine parts called for on the prints. General fabrication techniques were taught on the shop floor by the more experienced fabricators. I was taught by people working in the trade who had learned the skills from the people before them. This book contains some of those methods and is meant for use by anyone who builds with steel, including students in welding schools and trade programs, journeymen or apprentices in the steel trades, and people who might occasionally need the skills such as home mechanics, farmers, or maintenance people. This book assumes that you already know how to weld but doesn't require knowledge of how to read welding symbols.

Since steel is readily available and reasonably easy to work with using available tools, it is the obvious metal for manufacturing many useful items. It can be cut, formed, and welded into strong, tough, and relatively inexpensive items for industrial, farm, or domestic use. It is important to remember that there are varieties of steel which have special properties for specific applications and require special welding procedures. Covering all the different types of steel or metals other than steel is beyond the scope of this book, but the fabrication techniques will apply in most cases although the specific tools required may differ. I will go into methods for making sure that the final product comes out the way it was designed and I will try in a limited way to cover the mathematics involved in calculations necessary to the fabrication of steel.

I learned how to weld so that I could do metal sculpture and continued because it was a good way to make a living. I went through several welding schools and worked as a journeyman steel fabricator for over fifteen years building everything from security guards to offshore oil platforms and vibrating conveyors. I majored in math in college and have found the mathematics of triangles to be very useful in fabrication. I have successfully taught these skills to other fabricators on the shop floor. I have always found great pleasure in working with both my mind and my hands and found fabricating steel to be a challenge most of the time. I will attempt here to pass on the information I have accumulated to make it a little easier for you.

THE BASICS

THE MATERIAL

When we talk of steel we are actually talking about a wide variety of materials whose main constituent is iron. These materials vary widely in physical properties such as hardness, toughness, response to heat treatment, and weldability. These properties must be taken into consideration whenever steel is used as a fabrication material. Mild steel is by far the most common steel used and is easy to work with. It can be cut using an oxyacetylene torch and can be drilled, formed and welded with minimum consideration to protecting its physical properties. Under normal working conditions it can be arc welded with no preheat using a great variety of electrodes. It can be heated to high temperatures for bending or other forming and allowed to cool slowly or quenched in water with virtually no change in its physical properties. This is not the case with many other steels. Medium carbon steels, high strength-low alloy steels and high carbon steels are more temperamental. These steels usually require some preheat before welding and require special welding wires, fluxes, shielding gases, and techniques. Heat will affect the properties of these steels as will the rate of cooling. If you weld a high carbon or alloy steel with a welding electrode which has a coating high in hydrogen, such as E6010 or E6011, you are asking for cracking problems. Likewise if you heat up a high carbon steel you will change its physical properties, possibly in a disastrous way.

Since there are a fairly large number of types of steels available it is important to know what you are dealing with before you go too far. If you are working in a commercial shop, technical information can be gotten from the shop foreman or the engineering department. If you are working on your own, there are a number of sources of information from which you can get the information you need. The steel supplier should have information on the steels he sells. Lincoln Electric Company, which manufactures welding machines and supplies, has an excellent series of books on all phases of welding. *The Procedure Handbook of Arc Welding* and *Metals And How To Weld Them* are just two of their many titles.

SAFETY

Your safety and the safety of others around you is a primary concern in any shop. Hearing protection is essential since the noise in a metal working shop can be loud enough to permanently damage your hearing, particularly in the high ranges. Grinders and hammering have a long-range effect which is not reversible, but which is preventable. Every time you grind without hearing protection you loose a little more of your ability to hear high sounds. There are several different types of hearing protection ranging from complete ear covers to foam plastic cylinders which are compressed and then put into the ear where they expand and block out the sound. These are inexpensive and readily available and are meant to be used and then thrown away. Likewise, eye protection is essential. Grinders and other cutting tools throw pieces of abrasive and metal into the air. This material may come from something you are doing or someone close to you may inadvertently hit you with this material. In any case you don't want it in your eyes. Safety glasses come in types which look like reading glasses or may come as full face shields. They may be clear or tinted and are always made of material which can take impact without shattering. Safety glasses also reflect a certain amount of light and can help prevent arc burn from people welding near you. I have seen fabricators who routinely use the cutting torch without any eye protection at all. This not only leaves them unprotected from hot material which might come off the steel but also allows infrared light from the white hot metal to come into their eyes. This light has been shown to cause cataracts with prolonged exposure. Besides, the light is so bright that it is difficult to make a good cut because of the glare. Always wear cutting glasses or safety glasses with dark lenses when using the oxyacetylene torch. Steel toed boots, long sleeved shirts, gloves, hard hats if you are in danger from above, leathers to deflect hot welding slag, heat reflecting glove covers, and similar safety equipment are all designed to protect you from the environment you are working in. Take advantage of them. If there are lots of air born particles from paint or welding smoke, it is worth it to wear a respirator. You don't want to discover a few years down the road that you've done irreversible harm to your lungs.

Another aspect of safety has to do with the methods used to move material around. Extreme caution must be used when lifting material and it should always be assumed that the material could fall at any time. Rigging, the use of chains, clamps, magnets, hooks, and the like to move or position material, is best learned from someone with experience. People have been killed when chain hooks came loose because the chains were not put on correctly or when material fell out of a plate clamp or off of a magnet. Never pass material over someone without first warning them and never stand under a load. If you are using a crane to position a piece and need to be able to adjust it as it comes down, clamp a rope to it so you can stand away from it rather than holding on to it with your hand. Use common sense and don't be afraid to take a few minutes to figure it out. If you have any doubts, ask someone for advice or help. I can't emphasize this enough. It never appears foolish to ask for experienced help. Too many people have been killed or seriously injured when a load came loose and fell on them or someone else. **BE CAREFUL!**

At one point in my fabricating career we were building large marine cranes. The turrets stood about eight feet high and were about eight feet across. They were made of heavy plate some of which was as much as four inches thick. The turrets were not only heavy but they were hard to roll onto their sides for welding because the center of gravity was very low. If the turret was rigged correctly to the large overhead crane it could be picked up and would tilt until it would almost, but not quite, tilt over so it could be let down onto its side. The smaller jib crane was used to pick up the base a little so that the turret would lay on its side when the large crane let it down. It was done this way because the turrets were so heavy that the jib crane couldn't lift the whole thing but could take enough of the load to get them over the center of gravity. We had one welder who always knew how to do everything. He had never rolled one of these turrets and didn't feel he needed any advice either. The whole shop came to a halt when the turret he was working on crashed to the ground. His theory had been that he would pick it up low with the overhead crane and break its fall by using the jib crane attached to the top. He was standing about a foot from where it landed after the jib crane twisted like a pretzel when the falling turret pulled the end down. Needless to say this didn't make a lot of points with the boss and it wasn't long before he was "down the road." Don't be afraid to ask questions. It could save your life — or your job.

STRUCTURAL FORMS

Steel comes in a wide variety of forms as well as compositions. Getting the right steel from the steel rack or from the steel company requires knowing how the different forms are designated. This discussion will be brief, covering the most common forms only. Any steel company can supply you with a small book which describes how the different forms are measured and distinguished. There is a brief table of steel shapes and specifications in the appendix.

PLATE (PL) AND FLATBAR (FB)

Plate and flatbar are designated by giving the three dimensions of thickness, length, and width. For example pl. $\frac{1}{2}$" x48"×96" and pl. ⅜"x15"×34" are typical plate dimensions. ¼"x3"×5' Lg (Lg means long) and ⅜"x2"×37" Lg are typical flatbar dimensions.

ANGLE IRON (∠)

Angle iron is designated by the length of the legs and the thickness and then by the length. For example we could denote an angle iron as ∠ 3"×2"× ¼" ×5'-0". This is an angle iron with unequal legs which is one quarter inch thick and five feet long.

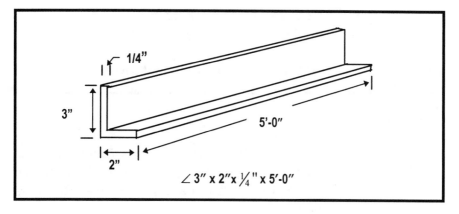

∠ 3" x 2"x ¼" x 5'-0"

WIDE FLANGE (WF)

Wide flange is structural material of the shape shown. It is measured in terms of its height and its weight per linear foot. Thus a WF8"@10# (10# means 10 pounds per linear foot) is different from a WF8"@21# even though the height of both is eight inches. It is important to keep this in mind since the specifications for a particular job might call for WF8"@10# and you might have to get the material from a rack where nothing is marked for size. You can't just take any eight inch wide flange you find. The way to determine what you are looking for is to look in your steel handbook (or the table at the end of this book) and check the other variables listed in it. These are web thickness and flange width. In this case a WF8"@10# has a web thickness of .170" and a flange width of 3.94" while a WF8"@21# has a web thickness of .230" and a flange width of 5.250". Don't let the fact that the dimensions are in decimals bother you. I have included information on converting decimals to fractions in the "Figuring it Out" section so that you can use your tape measure to make sure you have the correct material.

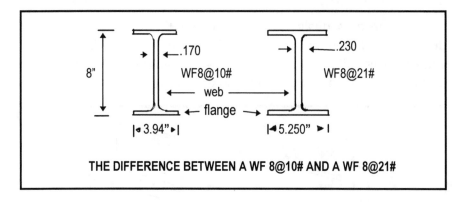

THE DIFFERENCE BETWEEN A WF 8@10# AND A WF 8@21#

CHANNEL

Steel channel is measured the same way as wide flange. The depth and the weight per linear foot are given. Again there may be several different weights for one depth and it is the web thickness and flange width that vary.

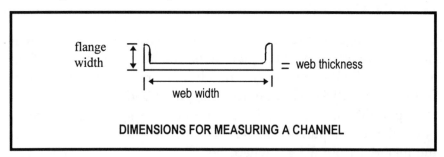

DIMENSIONS FOR MEASURING A CHANNEL

I recommend that you get a steel handbook from your steel supplier or welding supply store. It will cover all the various structural forms with dimensions, weights, and other technical information. With this information, you can determine what material to use as well as the total weight of a fabricated structure. Whether you are working for someone else or for yourself, knowing your material is of prime importance.

THE TOOLS

Before you can get very far in fabricating with steel you have to have some knowledge of the tools used in working with the material. Many of these tools are common to other trades and, as is true with every trade, there are a variety of tools which have been developed specifically for the metal working trades. In this section I will describe a variety of tools and how they are used.

LAY-OUT TOOLS

As you build with steel you will need to mark the material for locating holes, flanges stiffeners and the like. There are a variety of tools available to doing this lay-out. I will cover the most basic ones here.

MARKING TOOLS

The first tool you will need is some means of marking steel for lay out purposes. Since steel is generally bluish to reddish in color you will need a marking color which contrasts so that you can see what you are doing. You will often want to draw a line and then cut the metal along the line with a cutting torch, so you will need a mark which will be visible through welding lenses and which will not burn or blow away. The most common way of marking is with a soapstone, commonly called a chalk. Soapstone comes in pieces about five inches long by one half inch wide by either $\frac{1}{4}$ or $\frac{1}{16}$ inch thick. Soapstone also comes in $\frac{1}{4}$ inch diameter round pieces which can be sharpened like a pencil. Soapstone is brittle and is usually placed in metal holders which can be slipped into the pocket when not in use. This type of chalk makes a clean white line which can easily be followed with a torch. Silver pencils such as Berol Verithin #753 also work very well and can be used for fine marking because they can be sharpened like any other pencil. These pencils can be bought at many welding supply stores as well as at office supply stores. As with the soapstone this line does not burn away but is a little harder to follow because it is thinner. A scribe, which is a pointed piece of tungsten carbide in a pencil-sized holder, is used to make a scratch in the surface of the metal. This is usually reserved for precision lay-out work because of the accuracy which is possible. When long straight lines are needed a chalk line is usually used. This tool consists of a roll of string which is enclosed inside a container holding colored chalk.

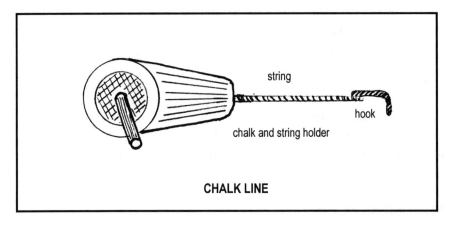

string

hook

chalk and string holder

CHALK LINE

As the string is reeled out it is covered with the chalk. The string is pulled taut and held down at each end on the desired line and then "popped" to release the chalk onto the surface. These are commonly used in the carpentry trades and the usual colors of chalk available are red and blue which show up quite well on wood but which are all but invisible on steel. Yellow is available at welding supply stores and shows up very well on steel. Chalk of this type has a tendency to burn and blow away from under the torch tip, so if you need to cut along the marked line it is best to go over it with the soap stone. There are a number of liquid markers which are used in much the same way as a pen to write on steel. These make a wide, semi-permanent mark and are generally used to write instructions such as welding symbols or machinery locations on the steel.

TAPES

Tapes come in various widths and lengths. I find that a 25 foot locking tape and a 100 foot reel tape do everything that I need to do. I prefer the locking tape because it allows me to pull the tape out to the desired length and lock it there while I work with both hands.

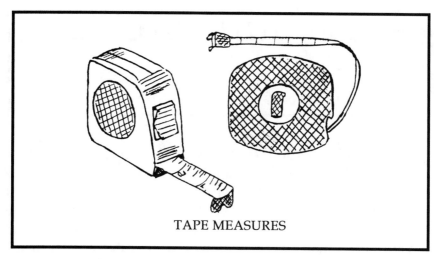

TAPE MEASURES

There are several things to be cautious about with a tape. First of all be careful how you use it to mark points. You will need to be able to come back later and know the exact point that you have marked. The simplest way is to make a mark like a V with the point being the location you want to mark. The figure below shows a distance of $3\frac{3}{4}$"

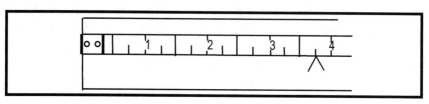

You will find that laying out points from the end of the tape can be inconvenient because the hook on the tape makes it stand up off the surface. You can put the one inch mark at the point you want to start from and avoid that problem. For example, if you want to measure 3¾"from a given marked point you can put the one inch mark on the given point and go an additional 3¾" to the 4¾" mark on the tape.

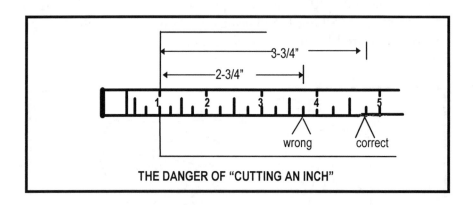

THE DANGER OF "CUTTING AN INCH"

BE CAREFUL not to measure to the 3¾" mark because then you have really only measured 2¾" since you started at the one inch mark. This missing inch is obvious on very short measurements, but it won't be that obvious when you are measuring 10'-3¾". "Cutting an inch" can lead to bad measurements, and everyone I know who fabricates steel has made this mistake at least once. Always double-check yourself. **Double checking always takes less time than reworking the job after its been built.** Cutting a foot, starting on the one foot mark, is usually a better practice because you can use the inch markers as they are and in <u>short</u> distances it is hard not to see a foot, but be careful anyway. The real skill in being a metal fabricator is doing it right the first time, before it is all welded together. When double checking yourself, try to use different reference points and cross check everything. It is far easier to go back and erase all your marks and start over than it is to break off all the pieces you've tacked or welded in place, clean up the surfaces and then start over again.

SQUARES

Squares are tools which are used to make right angles and come in several sizes and types for different applications. Framing squares are simply right angles made of metal for laying out perpendicular lines. They are marked in inches along the legs of the angle.

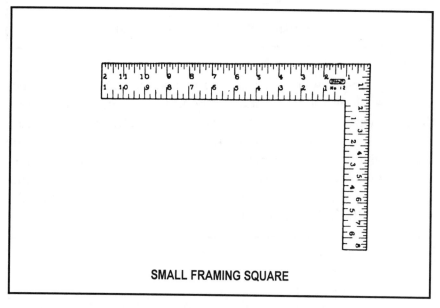

SMALL FRAMING SQUARE

Note that the numbers on the inside edge start at the inside corner and the numbers on the outside edge start at the outside corner. Not realizing this can lead to mistakes similar to "cutting an inch" with a tape measure. These squares are used for marking lines across a beam or for laying out perpendicular or parallel lines. They can also be used simply as a straight edge for drawing any line necessary. Quite often, when fabricating a structure, this square will be used to make sure that one item is perpendicular to another. I usually grind a small radius on the outside corner of my square since this lets the square fit over a tack weld or into a formed corner.

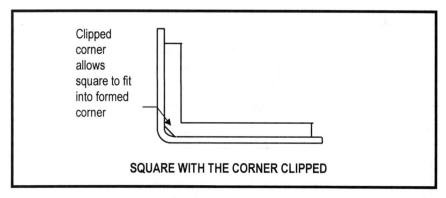

Clipped corner allows square to fit into formed corner

SQUARE WITH THE CORNER CLIPPED

The combination square is an adjustable square which is very useful in lay-out work. It can be adjusted so that the square lies anywhere along the blade and can be used to fit into tighter areas than a large, fixed square. Say, for example, that you need to draw a line which is two inches from the edge of a beam. Position the blade so that a little under two inches sticks out. This "little under" is to take into account the width of the marking tool so that the line comes out two inches from the edge of the beam. Then slide the square and the marker along the edge.

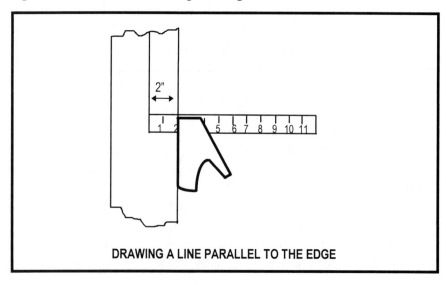

DRAWING A LINE PARALLEL TO THE EDGE

Another use is to mark a position on the web of a wide flange beam. That little curve where the flange and the web meet makes it difficult to do this accurately with a tape unless we happen to be at the very end of the beam. However, we can use a framing square and a combination square to mark this line quite accurately. The combination square has been adjusted so that the blade just reaches the web as shown in the drawing.

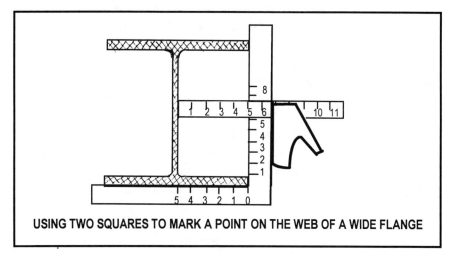

USING TWO SQUARES TO MARK A POINT ON THE WEB OF A WIDE FLANGE

Combination squares also have a 45° angle built into them which is very handy for mitering corners.

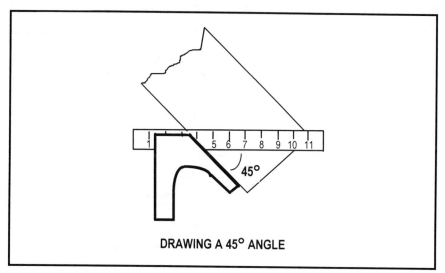

DRAWING A 45° ANGLE

Another type of square has a strong magnet on both legs and is useful because it holds two pieces at a right angle while you work on them or tack them in place.

STRING AND CLAMPS

A long string, say 50 feet, and a set of clamps can be used in different ways. I will give an example of one use although there are many others that will occur in day to day fabrication. Suppose you want to splice together two beams and that the resulting beam has to be straight in all directions. The beams can be put on four adjustable horses (I will cover making these horses later) to make things easier or they can be set on a flat surface and shimmed.

Use a tape or a square to mark the center of each beam on the top at each end as accurately as possible and then set up the pieces end to end as accurately as you can by eye.

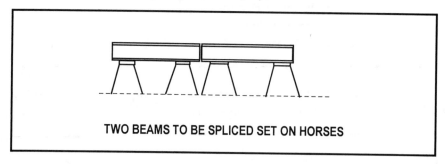

TWO BEAMS TO BE SPLICED SET ON HORSES

Now attach your string to the end of the first beam with a clamp and pull it tight to the end of the second beam clamping it down there also.

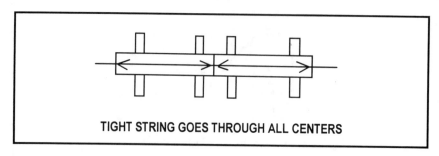

TIGHT STRING GOES THROUGH ALL CENTERS

The objective is to get the string to go through all of the points that we have marked so that the center line of the beams is lined up as shown in the drawing above. This will assure that the spliced beam is straight in one direction; but it is still possible that the beam will be bowed in the other direction.

To straighten the beam in this other dimension first insert small spacers of equal thickness under the string at each end. I usually use a couple of welding rods of the same thickness. Make sure the string is pulled very tight so that the amount that it drops in the middle due to the weight of the string is a small as possible.

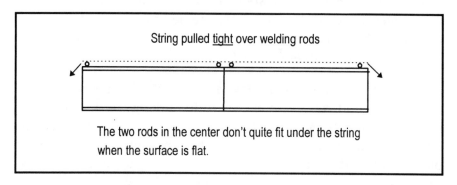

String pulled <u>tight</u> over welding rods

The two rods in the center don't quite fit under the string when the surface is flat.

Look at the splice point of the beams and see how far the string is from the surface at that point. If a string is pulled quite tight it will be about $\frac{1}{32}$ to $\frac{1}{16}$ inch lower in the middle than at the ends in a 30 foot span. When the beams are flat you should be able to take another welding rod of the same diameter as the ones at the ends and not quite slide it under at the splice point. If it is too close or too far then adjust the horses or shim the beams. The string method is very accurate if the string is drawn tight enough and it is generally more accurate than using a level, which only measures the surface it is sitting on. A string can also be drawn tight along an edge to check for straightness.

LEVELS

Levels are very useful tools in steel work, and they come in a variety of types which all have their usefulness in different situations. A two or three foot level is virtually indispensable in building frameworks of any kind.

A small bullet level with a magnetic base allows you to stick the level to the steel surface and still have two hands free to work. If you are attaching a vertical piece to a horizontal piece you can attach the magnetic level to the vertical piece and use one hand to hold it in place while using the other hand to tack weld the piece. Don't rely on a level for more than it can do. As mentioned before a level only measures the surface on which it sits. Also, the level depends on lining up a bubble within a pair of marks on the leveling tube. For short distances this is fairly accurate but over long distances it can lead to considerable error. A surface which rises 1° instead of being horizontal will rise $\frac{7}{32}$ inch in one foot. This turns out to be over 4" in 20'!

I use two other magnetic levels quite often. One is called an angle finder and consists of a dial with markings in degrees and a pointer which always points straight up. The other has an adjustable bubble which can be set to any angle desired. In the picture below these two types of level are use to set a 17° angle. With the angle finder the pointer points at 17. With the adjustable level the protractor is set at 17° and the angle is set when the bubble is between the lines.

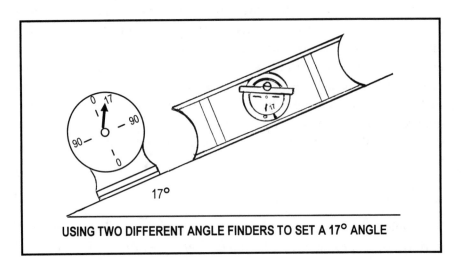

USING TWO DIFFERENT ANGLE FINDERS TO SET A 17° ANGLE

PROTRACTOR, COMPASS, AND PLUMB BOB

Protractors are used for laying out angles, especially on flat work. I use a clear plastic one that they sell for students because I find it handy to be able to see through it.

Compasses are used for drawing circles. Beside the normal dime-store type which work well with either a silver pencil or a round soapstone there are compasses available for drawing large circles. Your local welding supply store may have a large compass which uses a flat soapstone and which can be used to draw circles up to about four feet in diameter. Larger compasses are made by bolting a piece with a point attached to one end of a bar or rod and a piece to hold a drawing instrument such as a pencil, chalk or carbide point at the desired distance from the point. The only limit to the radius you can use is the length of the rod that you use to make the compass.

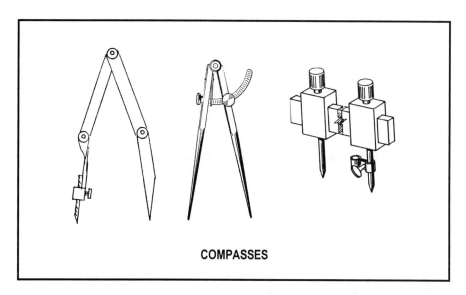

COMPASSES

A plumb bob is simply a pointed weight suspended on a string. Gravity pulls the weight straight down and you can take advantage of this to line things up. Suppose you have a piece which must be mounted directly below an existing piece. You can use the plumb bob to line up the pieces.

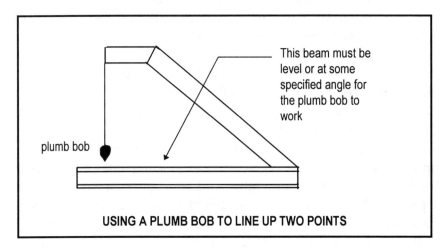

This beam must be level or at some specified angle for the plumb bob to work

plumb bob

USING A PLUMB BOB TO LINE UP TWO POINTS

The plumb bob is one of those simple tools, like the string and clamps, that can be used in a wide variety of ways. Quite often you will find that these simple tools can be made to do the job of much more expensive ones if you use your imagination. The chalk container for a chalk line is usually pointed on the bottom so that it can be used as a plumb bob if needed.

POWER TOOLS

Power tools are a very important part of the metal working trades. There are so many different kinds of electric and air powered tools that there are a number of books devoted entirely to that subject. I will not try here to go into any detail about most of these tools. Suffice it to say that power tools can be great time savers and that some projects in metal would be impossible without them. It is important to recognize the fact that the power which makes the work easier can also be dangerous to the user. A well equipped metal working shop will have a metal cutting band saw, a drill press, and possibly an iron worker for clipping corners and punching holes. There may also be a roll, a lathe, a milling machine, a brake, a shear, and several cranes for moving heavy material.

Angle grinders come in several sizes and are high speed grinders capable of removing steel quite rapidly. They are also capable of cutting the operator and the sparks and metal powder which they throw can set things on fire or injure other people in the immediate area. Make yourself familiar with any tools you will use and know the required safety procedures. Think ahead about what you will do if something unforeseen happens. Look around you to see if someone else could be hurt by what you are doing. Never assume that the machinery you are using couldn't break with catastrophic results. Chains can break dropping loads on you or someone else, grinding wheels can explode, hand held drills can catch and try to spin you instead of the bit with remarkable torque. Be alert and think ahead and power tools are a great friend. If you are careless they can be vicious when you least expect it.

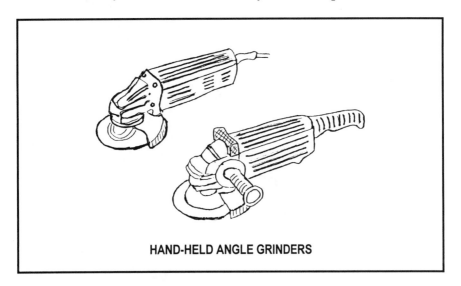

HAND-HELD ANGLE GRINDERS

YOUR EYE AND BRAIN

I've put this section last, but beyond all the other tools these are your most indispensable. With these you can figure out how to make the other tools do what you want them to do. Don't be afraid to come up with ingenious new uses for your tools. Finally, after you've used all your tools and double checked all your measurements, look at your work with a critical eye. You can <u>see</u> if lines aren't parallel or perpendicular to a very fine degree of accuracy. If you're building a frame which has to be flat, look across it to see if it really is. If a beam is bent it can throw a level off, but you will catch it with your eye if you look carefully. Remember, the mark of a good fabricator is knowing the product is right when it's done.

PUTTING IT TOGETHER

After you've done all the lay-out, all the holes are drilled, and all the other basic work has been done to prepare the pieces, you still have to put it together.

MAKING IT FIT

This section is really an extension of the previous one in that I will continue to cover the tools of the trade. Here, however, I will talk about them in a more specific context. Since steel is a very tough substance getting things to go together the way you want them to can sometimes be difficult. There are methods for forcing things into place and for holding them in place while you work on them.

HAMMERS

A common expression among steel fitters when something doesn't fit is, "Get a bigger hammer!" To a certain extent this is true and a fabricator will want several hammers of various weights and types. I use a one pound and a two and a half pound cross-peen hammer and a six pound sledge with a fourteen inch handle and have found that they suit my needs. The cross-peen hammer has a wedge shape for one of the working faces which is good for reaching into corners and which concentrates the force of the blow into a smaller area when needed. The six pound sledge is good for moving heavy objects or for exerting a large amount of force if necessary. For example, if you need to line up two heavy beams side by side you can use the six pound hammer to hit the end of one of the beams to move it into place. Steel can often be moved small distances with a hammer when it is far too heavy for you to move by hand.

PRYBARS

You will find that steel is rarely perfectly straight. Beams are often bowed and pieces which have been burned out using a torch are often warped by the heat involved. The fabricator's job is to force the steel into the right place and to weld it there so it will stay. This requires pushing and pulling and sometimes it requires a lot more power than you can get simply by hitting it. There are ways to multiply your power, however. These usually take advantage of inclined planes or levers. You will find that a good prybar can do lots of things. I have one made of $\frac{3}{4}$ " thick steel which is great for picking up the end of a beam or prying things apart. The long taper makes it easy to slide under a heavy object. Another useful small bar is round and pointed and can be used to line up holes that don't line up properly. This tool is often called a "spud" and may have a wrench or other tool at the other end.

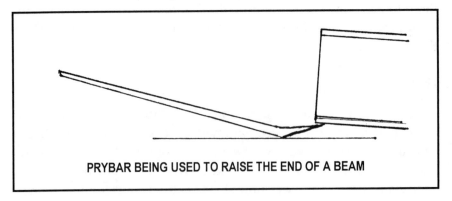

PRYBAR BEING USED TO RAISE THE END OF A BEAM

CLAMPS

Clamps are a common type of inclined plane. The threads go gradually up and give you the mechanical advantage you need as you tighten them. C-clamps come in a variety of sizes. Get good strong ones because they take a lot of abuse in a steel shop.

C-clamps can be used to pull pieces into alignment or for holding pieces together. In the figure below it is desired to line up pieces A and B. By using the clamp with a piece of steel the two pieces are drawn together as the clamp is closed.

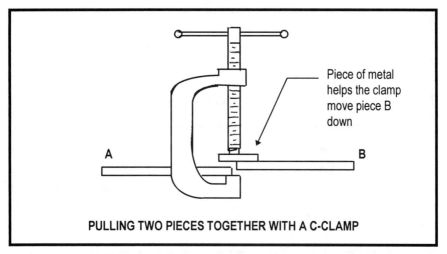

PULLING TWO PIECES TOGETHER WITH A C-CLAMP

Bar clamps, also called pipe clamps, are very useful. These are simply a two piece clamp which uses a piece of pipe or a steel bar as the connecting arm. By using different lengths of pipe you can make clamps to meet your particular needs. The clamps most commonly available are designed as furniture clamps and are not intended to exert the forces commonly applied to them in metal shops, so don't be surprised if these disintegrate over the course of time. There are special, heavy duty, clamps made for steel work but these are not as readily available as furniture clamps.

Bar clamps are usually used to pull two pieces together for fine tuning the dimensions required. The figure below shows how they can be used for pushing two pieces apart by reversing the moveable clamp and screwing the adjustable clamp all the way in. By unscrewing the clamp the pieces are forced apart.

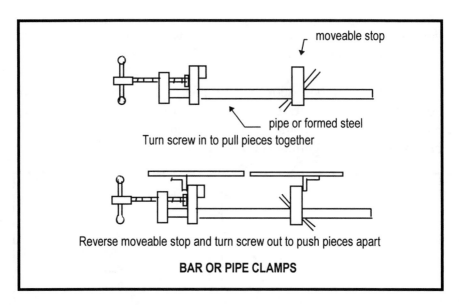

moveable stop

pipe or formed steel

Turn screw in to pull pieces together

Reverse moveable stop and turn screw out to push pieces apart

BAR OR PIPE CLAMPS

Adjustable clamps are used to hold pieces in place while they are being worked on. The adjusting screw in the handle determines the opening of the jaws and the pressure with which they hold. These clamps are easy to use and are indispensable for everyday fabrication use.

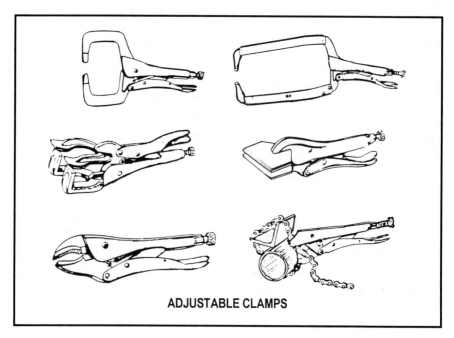

ADJUSTABLE CLAMPS

JACKS AND COME-ALONGS

Jacks, either hydraulic or mechanical, can be of great use when heavy items need to be lifted for leveling or when pieces need to be spread apart. They range in size from a couple of inches long up to quite large jacks for heavy work, with capacities from about 1000 pounds to many tons. Some hydraulic jack cylinders use a hydraulic pump attached with a hydraulic line so that the cylinder can be put in place and then be operated from another location. This can be important if the situation is dangerous or, more commonly, where the area is just too tight to work in.

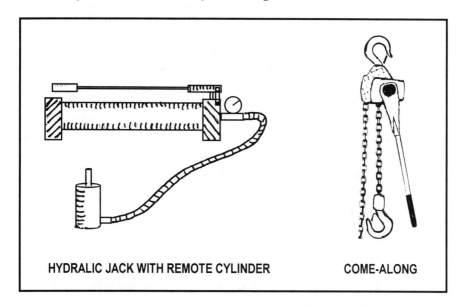

HYDRALIC JACK WITH REMOTE CYLINDER   COME-ALONG

Come-Alongs are chain pulls or hoists which range from one half ton up to about three tons in capacity. Unless you are doing very light work the cable type of come-along won't be adequate, because they usually have a ratchet and pawl action and can be difficult to release easily and safely. Chain pulls have clutches which are engaged for either tightening or loosening the chain. This allows you to tighten up a little at a time or slack off a little without having to go to the next stop. This infinite adjustability can be important.

DOGS AND WEDGES

As a last resort, when things really don't want to move and your clamps, jacks, and prybars aren't strong enough, you still have another option left—wedges and dogs. Wedges are pieces of steel cut as is shown below.

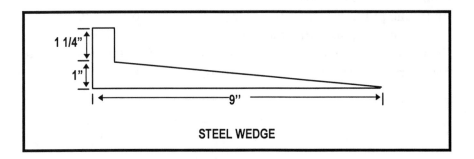

STEEL WEDGE

The dimensions shown are typical for shop wedges although it is not uncommon to find them twice this size. The important thing is that the taper must be gradual so that they can be driven into place without slipping loose. The raised end is there so that you can remove the wedge with a hammer or another wedge when you are finished. They are usually made from three quarters to one inch thick steel; but they can be made of anything available which will do the job.

A dog is any piece of steel that can be used with a wedge to get the desired result. They are usually made of scrap when the need arises and thrown in the scrap when you're done. There are a couple of configurations which you might want to keep in your tool box but they are so easily made it isn't necessary. The dog can be made of heavy material and welded down and the wedge can be driven with a large hammer so that a large force can be exerted. Usually you only need to tack the dog on the side from which the wedge will be driven which makes it easy to remove the dog later by hitting it with a hammer instead of having to resort to the grinder or the torch. The picture below shows two examples of a dog tacked to one piece of steel while a wedge will be used to force the second piece of steel into place.

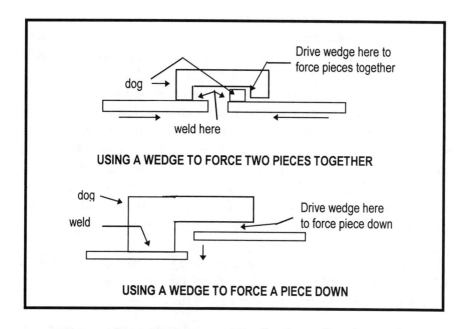

USING A WEDGE TO FORCE TWO PIECES TOGETHER

USING A WEDGE TO FORCE A PIECE DOWN

Dogs can be used to push things apart as well as to pull them together Let your situation be your guide. Use your imagination to come up with a tool that will do the job for you. Sometimes the best thing to do is, "Get a bigger hammer!"

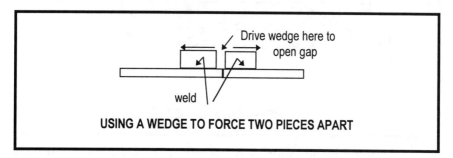

USING A WEDGE TO FORCE TWO PIECES APART

FIT-UP: LEVELING AND SQUARING

For much of the work you will be doing it will be convenient if you can work at a comfortable height. Working on the floor may be necessary for some things, but if you can work in a sitting or standing position you will appreciate it at the end of the day. Steel tables are easily made and are convenient because they can be used as part of the welding circuit for arc welding. Pieces can be welded to the table to hold them during fabrication and then cut loose and the welds ground off so that the table remains smooth. The table should be made at a good working height so that you don't have to stoop over to work. I like to use a piece of ¾ or 1 inch thick plate since it will not deform when it is welded or hammered on. A common mistake when building a table is to forget that heating steel causes it to warp toward the heated side. Excess welding on the legs can cause the corners of the table to warp down and stiffeners welded on the underside can cause the surface of the table to warp. Having a nice flat surface to work on can be important so make the table strong but avoid overwelding.

For a lot of fabrication you will need to have moveable supports which are adjustable for leveling. Steel horses are handy and can be made by using the figure below as a guide.

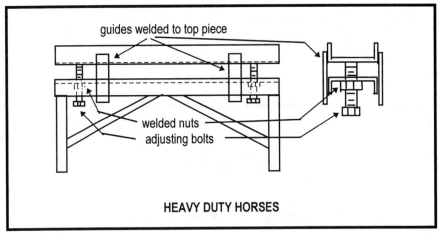

HEAVY DUTY HORSES

They should be sturdily built since they commonly carry quite a bit of weight. The nuts are welded over holes cut in the bottom piece so that the bolts can be threaded through them to lift up the upper piece. Guides are welded onto the upper piece so that it will stay in position when raised by turning the bolts. With these horses you can level your working surface even if the floor isn't level, which most aren't.

Rectangular frames are very common in steel fabrication. It is usually important that the frame really be a rectangle with all four corners being right angles. If the frame is small a framing square is probably accurate enough, but for any large frame this is not adequate. A principle which is used over and over again is that the diagonals of any rectangle are of equal length. For example suppose you want to build a frame which is 10 feet long and 4 feet wide and this frame must be both flat and square. First, place the frame members on the adjustable horses and position them so that the beams are 4' apart to the outer edges and the ends are as square as you can get them by eye. Use the adjusting screws to raise or lower the ends of the beams so that they are level when checked with a long level. Next squat down and look across them to see if they are really in the same plane. While a level is pretty accurate you will find that your eye can spot whether two objects are in the same plane very accurately. Once that is all done you are ready to get the ends square.

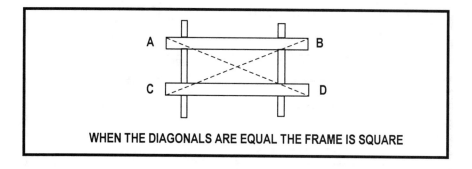

WHEN THE DIAGONALS ARE EQUAL THE FRAME IS SQUARE

Measure the length of the diagonal from A to D and record the measurement. Then measure from C to B and compare this length with the first. For the frame to be square the two measurements must be equal. Lets say that C-B is ¾″ longer than A-D. Take your hammer and tap end B back about ⅜″. Recheck the 4' dimensions since they may have changed and then recheck the diagonals. After repeating the process as often as is necessary, the frame will be square and level and you can proceed to put on the cross members or whatever.

LAY-OUT & MATERIAL PREPARATION

A lot of the work of a fabricator is marking and laying out material so that holes can be drilled or burned or so that gussets or other structural members can be added on. The fabricator should look over the project to determine which elements are critical and which can give a little. Steel coming to the fabricator may not be uniform. Lengths may not be exactly right and torch cut parts may be off in some way. It must be decided which surfaces are critical and which are not. For example, suppose we are given the following part to fabricate. We have a six inch wide-flange with holes drilled in it and gussets to strengthen it. Two one-inch plates with holes in them will be welded onto the wide-flange as shown.

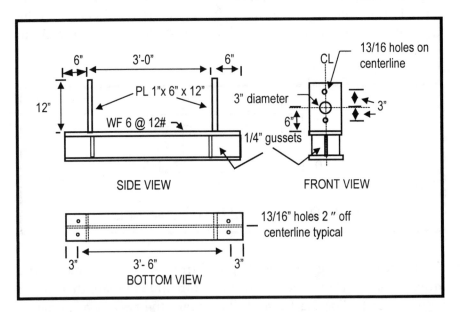

When you get the one inch thick plates you find that rather than being 6"×12" one is $6\frac{1}{8}$" x $11\frac{7}{8}$" and the other is $5\frac{7}{8}$" x $12\frac{1}{4}$" and the beam is 3'–$11\frac{3}{4}$" long instead of 4'–0" long. You could spend a lot of time on the plates making them the same or you could get a new plate and a new beam but more than likely you can use what you have.

CRITICAL SURFACES

First, determine whether the discrepancies will cause problems later on. If the 12 ¼" plate is supposed to fit into a 12 " space then it will have to be trimmed. For this example we'll use it as it is. To determine how you will lay out the holes you must decide which dimensions are critical and which aren't. In this case the distance from the beam to the holes in the plates is clearly marked so you will use that end of the plate as one of the reference lines. The holes in the beam should be symmetrical about the center line of the beam. The holes in the plate also lie on this centerline so you will use the center line of the plate as a second reference line. Your final layout of the two plates will look like this:

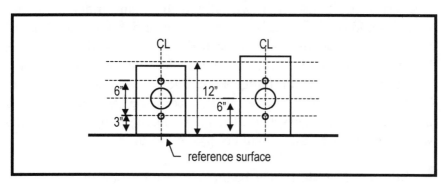

Note that the bottom edge and the center lines are now the reference lines.

Now you can lay out the beam. Determine if the fact that it is ¼" too short is critical. If it is, then get another beam; if not then make it work. It is possible that the beam butts up against something and that the dimensions from that end are critical in which case you would do your layout from that end. In this case we will assume that the plates should be symmetrical about the center of the beam. First, find the center of the beam. It is at half of 3'-11¾" which is 1'−11⅞". Then measure out 1'−9" in each direction from this line which gives two lines which are 3'−6" apart. Now measure the width of the beam at these two points and find the center of the beam in that direction. From those lines you lay out the holes which are 2″ off center line.

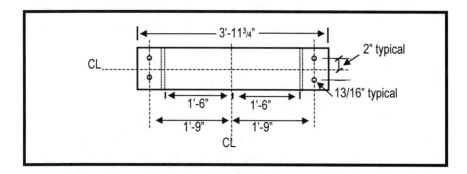

Using the centerline of the length of the beam, you also mark the location of the gussets that will stiffen the beam under the plates that you laid out previously. These plates and gussets are 3′ apart so you will measure 1′−6″ in each direction from the centerline. Use your combination square to transfer the location from the bottom flange to the top flange so that you have a reference line to locate the plates. You can also use the combination square to mark the location of the gussets on the inside of the flanges and on the web of the beam so that it will be easy to locate.

Quite often wide flange beams come with one of the flanges tipped down. If this is the case use the gussets to straighten the beam out before going on. Usually this can be done by driving the gussets in on the short side. If this isn't adequate, use a jack to spread the flanges apart and tack in the gussets. In any case tack weld the gussets in place on the outside of the lines you drew.

Before attaching the plates all the holes should be drilled and the center holes should be cut out to 3″ diameter as called for on the print. The final step is to put the plates on top of the beam. Find the centers of the lines that you have transferred to the top flange. Now match the center line of the plate to the center line of the beam being sure to put the plate on the outside of the line because the inside distance between the plates needs to be 3′−0″. Use a square to make sure the plate is perpendicular to the beam and tack weld the plate on one side. Check with the square to make sure the plate is perpendicular. It probably won't be since the tacks will have pulled it. When you've got it right tack the other side of the plate. Check again to make sure it is still perpendicular. If it isn't you may have to hit it with a hammer. Put in the second plate. After both plates are tacked in place double check yourself by measuring the distance between the plates at both the top and the bottom. If the distance is not 3' − 0″ find out why. Also double check the distance from the beam to the holes to make sure the plates aren't on upside down.

Welding distortion must always be taken into account by the fabricator. Before welding the plates into place they should be braced to keep the welding from pulling them out of square. Using some flat bar or scrap steel brace from the top of each plate to the beam and also between the tops of the plates. Position the braces so that they are easy to work around and easy to remove when the welding has cooled. If there are specifications concerning the welds be sure to write them on the piece to be welded using a steel marker. Quite often the fabricator will pass the piece on to a welder who may not have access to the print. If there are no welding symbols on the piece and the welder doesn't weld it correctly, it is the fabricator who is responsible.

JIGS AND FIXTURES

The procedure outlined above would be fine in a situation where there were only a few of these items to make on a one time basis. However, if this part were needed on a continuous basis or even of this were a one-time order for a large number of them you would alter the procedure to speed up production. First of all you could make a template for the plates. A template is a pattern which is used to transfer lay-out information without having to go through the process of measuring each piece individually. In this case it could be a piece of sheet metal with tabs bent on two perpendicular sides so that it can be simply positioned on the plate.

The template would have small holes drilled at the location of the holes to be transferred. Using small holes allows for transferring the location with a punch while maintaining accuracy of location. If the holes are too large it is hard to find the center with the punch. Although it may take longer to lay out and build the template than to lay out a single plate, if there are a lot of plates to do you will definitely save time. Time can also be saved by drilling more than one plate at a time. This is called stack drilling. Several plates can be stacked on top of each other and tack welded together. Be sure to use a square to line up the critical edges otherwise the holes will not be in the correct place on the lower plates. The top plate can then be laid out using the template and all of the plates drilled in one operation. It is also possible to build a "jig" which will position the plates on the beam so that individual layout and squaring can be eliminated. In the case we considered before, a jig might consist of a set of stops welded to a steel table for positioning the beam and two structures against which the plates can be clamped to hold the required dimensions and to hold the plates perpendicular. The welding can often be done in the jig while all the pieces are clamped in place which also eliminates bracing and rehandling.

TORCH WORK

In the job shop, where the fabricator is called upon to do a lot of one of a kind work, it is often necessary to make the pieces from scratch. The oxyacetylene cutting torch is a very important tool for this kind of work and it is important that you understand how to use it in a variety of applications. Knowing how to weld with a torch is of limited importance; but knowing how to cut steel with the torch is vital. Since this book is written on the assumption that you don't have much experience, I will lay out the basics here.

IMPORTANCE OF A CLEAN TIP

First of all, it's important that you have a good selection of cutting tips. You can't do a good job of cutting heavy plate with a small tip and cutting light gauge steel with a large tip is equally difficult. The instruction book that came with the torch will list steel thickness for each tip size along with gas pressures required. The end of a cutting tip has a center hole surrounded by a circle of small holes. A mixture of oxygen and acetylene comes out of the outer holes and the flames are used to heat the steel to near the melting point. High pressure oxygen comes out of the center hole when the cutting lever is pushed. When the steel is near the melting temperature this high pressure oxygen causes it to burn and also blows the resulting slag out of the way thereby cutting the metal. In order to make clean cuts it is necessary that the torch tip be kept "clean". As a torch is used bits of slag can get into the holes in the tip. This doesn't usually affect the outer holes much and they need little attention, but the high pressure hole requires periodic cleaning to remove any slag that obstructs the free flow of oxygen. The easiest way to tell when this is necessary is to light the torch and then press the oxygen lever. You should see a long thin stream coming from the end of the torch for a distance of a foot or more. If the stream is short and grows wide quickly the torch tip is dirty. After you use the torch awhile you will be able to hear the difference between a clean tip and a dirty one. A clean tip has a distinct crackling sound.

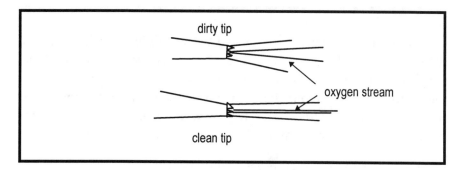

The oxygen stream is what cuts the steel. If it is long and thin the cuts will be clean with very little slag stuck to the metal on the bottom side. If it is short and wide the cut will be irregular and there will be big globs of slag stuck to the metal which will have to be chipped off. The other workers in the shop will tell you it "looks like an alligator bit it off." Good cuts take practice and a steady hand. The more you cut the better you will get.

TOOLS FOR CUTTING STRAIGHT LINES AND CIRCLES

There are several tools which greatly improve the quality of the cuts made with the torch. These are the burning bar, the circle cutter, and hole guides. The burning bar is an aluminum bar with magnets in it which is placed along the line to be cut. By moving the torch tip along the edge of the bar a straight cut is made. Note that to cut along a given line you have to take into account the width of the cutting tip when placing the burning bar.

Circle burners are adjustable compasses with an attachment on one end which fits over the torch tip.

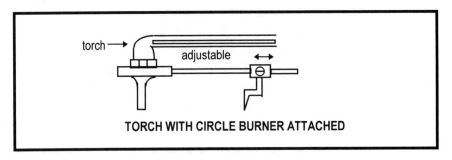

The center of the circle is marked with a center punch mark and the center hole of the torch tip is set to cut the desired sized circle. Starting a burned hole in a plate causes slag to blow back toward the cutting tip. To avoid this you can drill a small hole where you want to start the cut. This works well with the circle burner because it allows you to make the cut in one smooth motion.

GUIDES FOR BURNING SMALL HOLES

Hole guides are used for burning holes which are too small for the circle burner, that is for holes smaller than about 2" in diameter. Using these guides holes as small as $\frac{9}{16}$" can be accurately burned. To make a guide of the size needed proceed as follows:

1) Measure the distance across the end of the torch tip you wish to use. 2) The size of the guide will be the size of the hole you desire to burn plus the diameter of the tip. 3) Find a washer which is the correct size. 4) Weld a short piece of $\frac{1}{4}$" thick flat bar to one side of the washer making sure that you leave some distance between the end of the bar and the opening in the guide. This bar gives you something to clamp or tack in place while you burn the hole and also gives you clearance under the guide so that it is easier to use. 5) Use a piece of emery cloth to smooth the inside of the washer so the torch tip can slide easily without sticking.

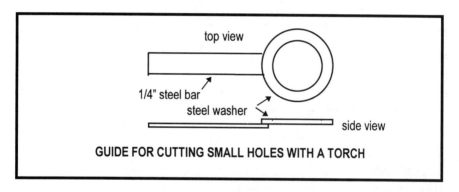

GUIDE FOR CUTTING SMALL HOLES WITH A TORCH

As an example, Victor tips are about $\frac{1}{4}$" across. If we want a guide for a $\frac{9}{16}$" hole we will need a $\frac{13}{16}$", ($\frac{9}{16} + \frac{4}{16}$) diameter guide. That is the washer size used for a $\frac{3}{4}$" diameter bolt. Weld on a short piece of $\frac{1}{4}$" flat bar and the guide looks like the drawing above. To use the guide, line it up with the position of the desired hole and clamp or tack it down. Start the cut in the center of the guide and then move it to the edge and run the tip around the inside of the guide. I usually drill a small ($\frac{1}{8}$" diameter) hole at the center of the desired hole in which to start the cut to eliminate the problem of slag blowing back up under the guide when the cut is started.

Beveling for welding can also be done with a torch. On light sections you will probably want to use a grinder to cut away the metal. On heavy sections this is impractical and the torch is the right tool to use. You can use the burning bar to insure a straight bevel or you can do it free hand if you are good with a torch. I've found that pre-heating the area to be beveled will help make the bevel smoother. The main mistake made in beveling is to cut away too much metal. The figure below shows a correct and an incorrect bevel. The bevel should be what is necessary to get a weld all the way through and no more. If the bevel is too great it requires more weld metal to fill it and requires more time, more cost, and causes more welding distortion.

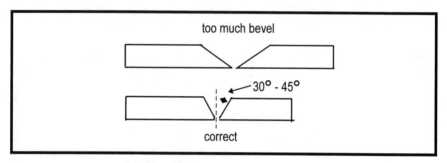

The common structural forms, angle iron, channel, wide flange etc., are widely used in steel fabrication. The fabricator will be called upon to cut these and to fit them into the structure and this must be done in such a way that the fit up is good and that welding can be performed easily. There are preferred ways of cutting to get the desired results. Usually the best way is to saw the pieces to the desired length and the desired angle. A saw is not always available or practical so I will cover procedures for using the torch to cut the material. Suppose you want to make a square cut. The first step is to lay out where the cut is to be made and then to mark it in such a way that the cut can be made easily. For angle and channel this is straight forward as shown below.

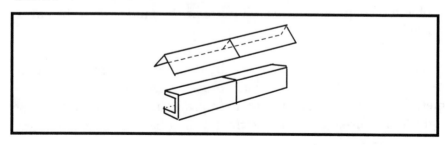

In each case start at an edge, cut to the corner and then finish the cut to the other edge. The torch will remove material in the process of cutting. Instead of cutting directly down the line, cut next to the line away from the piece you want to save. This assures that the piece will be the right length. If you want to cut a number of 12" long pieces out of one long piece you have to take into account the material lost with each cut. If we cut each piece on the line and loose $\frac{1}{8}$" at each cut, the length of the first piece will be $11\frac{15}{16}$" long and the other pieces will each be $11\frac{7}{8}$" long.

To avoid this lay the first piece out at 12" and the remaining pieces as $12\frac{1}{8}$". Cut on the outside of the line each time and all of the pieces will come out to be 12" long. Since the actual amount of metal removed may vary a little it is not recommended to lay out a large number of pieces at one time as any error will accumulate as you go along.

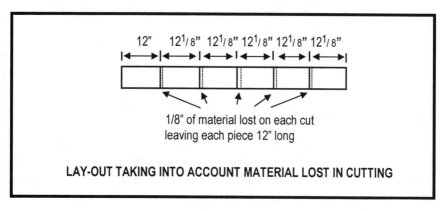

LAY-OUT TAKING INTO ACCOUNT MATERIAL LOST IN CUTTING

To make a square cut on wide flange proceed as follows. Lay out the cutting line on the top flange and then continue this line to the bottom flange using a square. On one side of the beam use the combination square to mark the inside surfaces of the flanges and then mark the web. Start your cut on the web and burn up as close to the flange as possible leaving a small round hole at each end. This hole will give the torch a place to come through when you are cutting the flanges and will make a much smoother cut.

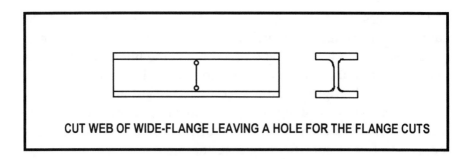

CUT WEB OF WIDE-FLANGE LEAVING A HOLE FOR THE FLANGE CUTS

Cutting structural steel at an angle is similar but requires some additional technique. I will give an example using a wide flange since it is fairly complex. Other structural forms will be treated similarly.

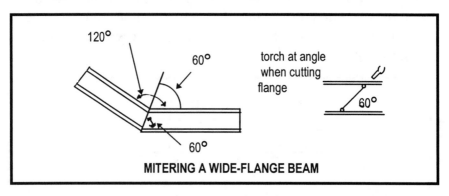

MITERING A WIDE-FLANGE BEAM

Here we have two wide flanges which are to meet as shown to make an angle of 120°. In order that the edges will be the same length both pieces must be cut at the same angle. This angle will always be half the total angle, in this case it is 60°. Laying out an angle can be done in a variety of ways depending on the tools available. If you have an adjustable square which can be set to any desired angle that is the easiest. A protractor and a ruler will also work. Later I will show you how to lay out any angle if you have a calculator. The object in laying out the cut is to draw a line all around the beam which will be what a saw cut would look like if you sawed the piece in two. First, mark one side of the beam with the required angle. Use a square to extend those marks across the bottom and top flanges of the beam and then use those lines to mark the angle on the other side. Use a combination square to mark the inside of the upper and lower flanges. Now connect the lines on the inside of the upper and lower flanges across the web. Start your torch cut on the web and cut to the flange leaving a small hole as before. Now cut the flanges. Remember, these are not square cuts. The torch must be held at an angle so that the cut comes from the top of the flange through to the cut in the web. You can use the line on the edge of the flange to help you set this angle.

JOINT DESIGN

There are lots of ways to join members together, but some choices are better than others from the point of view of time spent doing lay out, cutting, fitting and welding. For example, there are several ways to make a 90° angle out of angle iron. If you look at the two methods shown in the next drawing you see that they give the same result but that the approach is different. Which is preferred? From the point of view of lay out and fitting the one on the right is best. Only one piece is cut and the cut is square. This is easy to measure, lay out and cut. The other example has two lay outs which are 45° angles. The back corners have to be beveled. The two pieces have to be identical and the angles have to be right or they won't fit together correctly. From the welding standpoint they are about equivalent. The wise choice is the one which requires the least time. As you look through these possibilities keep in mind that you want the easiest, most efficient means of meeting the requirements of the problem. Extra cutting, grinding, and fitting take time.

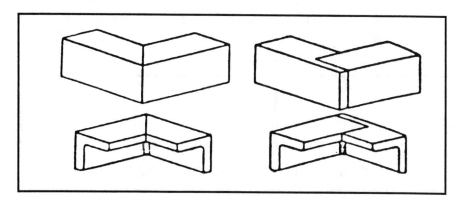

A common way of joining two wide flanges of the same size is shown below.

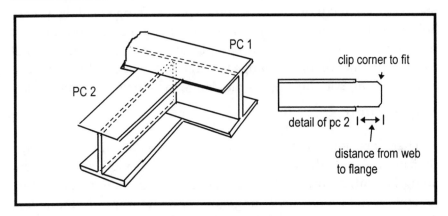

Measure the distance from the edge of the flange to the web. This can be done by laying a straight edge across the flanges and measuring to the web using a tape measure or your combination square. Mark the flange both outside and inside and across the web at the distance measured above. Mark the width of the web on both flanges and cut off the four pieces of flange leaving the web sticking out. Now mark the contour of the inside of the beam on the web and cut it to fit. If you have a thin slice of beam cut by the saw for a template it is easier. Finally trim the corners off so that it will fit inside of the other piece.

FLAME STRAIGHTENING

The heat involved in the welding process can cause metal to bend. When you tack weld something the tack tends to pull the item as it cools. If you weld two plates together on one side only the plates pull up on the side of the weld.

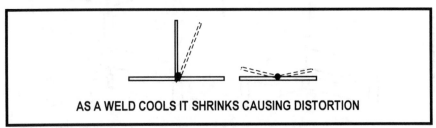

AS A WELD COOLS IT SHRINKS CAUSING DISTORTION

Mild steel is affected by heat in a predictable manner. When the steel is heated it expands and when it cools it contracts. This property can be very useful because it gives us the ability to straighten out distortions caused by welding. The key behind using this effectively is knowing where to heat and how to cool the steel to get the desired effect. The steel needs to be heated to red heat and can be cooled with a spray of water or with rags soaked in water. I will give an example which will show the basics. A frame I was building was made of two parallel eight inch wide flanges with six inch channels welded across them at about three foot intervals as shown below.

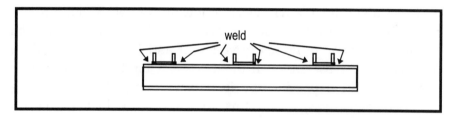

After welding, the frame came out looking like this due to welding distortion.

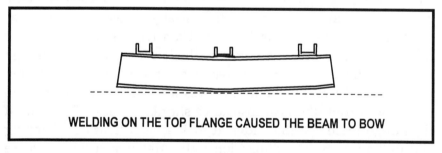

WELDING ON THE TOP FLANGE CAUSED THE BEAM TO BOW

Since the bending actually took place at each weld we need to heat in the same area but on the opposite flange. I heated an area an inch or so wide on the flange opposite the weld doing one weld at a time. After that area was hot, I heated a V-shaped area on the web from the weld to the lower heated area, as shown in the next drawing, and then cooled the whole area using wet rags. This caused the heated area to shrink thus removing the distortion of that weld. By doing this at each weld the beam was straightened out.

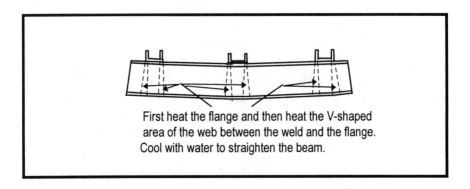

First heat the flange and then heat the V-shaped
area of the web between the weld and the flange.
Cool with water to straighten the beam.

This same principle will work in lots of different situations and you must be aware that you can overdo it. You should look each situation over carefully before you start and heat up as small an area as seems reasonable. If there is stress built into a structure, such as would happen if the item were clamped down during welding, heating may release this stress and cause the item to bend unpredictably. Other steels than mild steel can also be unpredictable due to built in stresses. In one shop where I worked a piece of T-1 steel got substituted for mild steel in a trough section. The outside dimension of this trough at the discharge end was rather critical and after welding and assembly it was $\frac{1}{4}$" wider than called for. I heated a line from the lower corner of the trough upward to the edge about two feet back and cooled it with water. Heating the T-1 released built in stresses which caused the steel to move away from the heat instead of shrinking back. This caused the trough to be $\frac{1}{2}$" too wide! It took several hours of clamping and pulling to get the dimension right. T-1 is a special purpose alloy steel and doesn't react like mild steel. The properties of some steels will be ruined if you treat them with heating and cooling this way so find out what you are working with and what the effect will be before you start.

FIGURING IT OUT

Anyone who gets seriously into metal fabrication will run into the need to figure out information which is not directly given. The blueprint may not give the information in a way which is easy for you to use or you may have a project of your own and need to calculate a length or angle. In those cases you will need to "translate" the given information into something you can use. My intention in this section is not to teach you detailed mathematics but to give you another tool which will help you do a better job. I suggest that you read the rest of the book with a pad of paper and a pencil so that you can work along with the examples. That way you'll understand what is being done and will be able to use it when you need it.

I recommend that you get a cheap scientific calculator similar to the one pictured on the next page. These are small enough to fit in your shirt pocket and contain all the information that you'll need for figuring out angles and lengths. These are the same methods that engineers use to calculate the dimensions given on blue prints and they are very useful in building things. The features that we'll use for this book are squares, square roots, SIN, COS, and TAN functions, and **arcsin, arccos,** and **arctan** functions. (Note that on some calculators the keys for **arcsin, arccos,** and **arctan** may be labeled as ASIN or SIN-1, ACOS or COS-1 and ATAN or TAN-1.) I recommend that you get a calculator which will do arithmetic with fractions since that makes doing calculations easier.

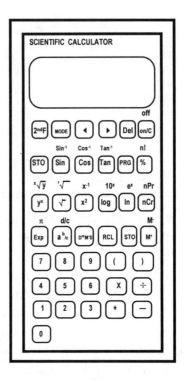

CHANGING DECIMALS TO FEET AND INCHES

Quite often, when we are calculating dimensions for fabrication, we will get answers which are decimals. Since the measuring tools we commonly use are in feet and inches we need to be able to convert to those units if we are to use the calculator effectively. What follows is the basic method; followed by a practical example.

CONVERTING DECIMALS TO INCHES AND FRACTIONS

1) Take the number which you wish to convert, set aside the whole number part and use the decimal part. For example, if the number is 5.357", set aside the 5 and use the .357.

2) Decide how accurate you want the result to be. Is an eighth of an inch adequate or should you use sixteenths or even sixty-fourths? I usually use sixteenths or thirty-seconds.

3) Multiply the decimal part by the size of the measuring unit you are looking for. If you want eighths of an inch then multiply by 8, if you want 16ths multiply by 16, if you want sixty-fourths, multiply by 64.

4) Round your result to the nearest whole number and that is the closest equivalent to the decimal.

6) Add back the whole number you took away in step 1 and that is the final measurement.

NOTE: The fraction you get is a close approximation of the decimal but it is usually not exact and may cause error if used repeatedly.

EXAMPLE

Suppose we have a 19' long beam and we want to divide it into 17 equal parts so we can put in gussets. First convert 19' to inches by multiplying by 12 to get 228". Now divide this by 17 to get 17 equal parts each of which is 13.412" long. What we need to know is the fraction of an inch that .412 represents. To get the result to the nearest sixteenth of an inch, set aside the 13" and multiply .412 by 16 to get 6.592 which rounds up to 7. So .412 is approximately equal to $\frac{7}{16}$". This will be written as .412 $\approx \frac{7}{16}$ (This means that the two numbers are approximately equal) Add back the 13" and you find that each part is ~13$\frac{7}{16}$" long. (The ~ before a number means it is approximate) Had we wanted the result in 32nds we would multiply .412 by 32 to get 13.18 which gives ~$\frac{13}{32}$"so each part would be ~13$\frac{13}{32}$" long. NOTE: Neither of these measurements is exact because we rounded off to the nearest whole number. This means that you will accumulate

50

a little error as you lay out the locations of 17 pieces. I would probably start from both ends of the beam and then make the ones in the middle fit as best I could. From now on I will show the calculations and write the result as something like $.412 = \frac{7}{16}$, meaning that this is the closest approximation to the unit of measure we are using.

CONVERTING MEASUREMENTS IN INCHES TO FEET

Lets convert 164.572 inches to feet and inches.
1) Divide 164.572 by 12 to get 13.7143'.
2) Set aside the 13' and multiply .7143 by 12 to convert it to inches. We use 12 because each inch is $\frac{1}{12}$ of a foot. The result is 8.572".
3) Set aside the 8 and multiply .572 by 16 to get 9.152. Rounding off we get $\frac{9}{16}$". Now add back the parts set aside to get $13'-8\frac{9}{16}$".

EXAMPLE

We could have solved the division of 19 feet into 17 equal pieces without converting to inches. We would have divided 19 by 17 to get 1.1176' and then converted .1176' to inches by multiplying by 12 to get 1.4118".and then converting .4118 to sixteenths by multiplying by 16. (.4118)(16) = 6.588[*] so we have about $\frac{7}{16}$". Now adding back the parts set aside we get $1'-1\frac{7}{16}$" which is the same as $13\frac{7}{16}$".

We usually want to convert decimals to fractions but occasionally we want to convert a fraction to a decimal. To do this, simply divide the numerator by the denominator. For example to convert $\frac{7}{16}$ to a decimal divide the numerator, 7, by the denominator, 16, to get 0.4375.

---

[*] (.4118)(16) means multiply the two numbers.

TRIANGLES

Triangles are the basis of a lot of the mathematics involved in fabrication. Solving problems involving triangles will help you find dimensions and angles to make your lay-out easier and more accurate. Being able to get answers on paper can also save you a lot of time and trouble. I'll give an example.

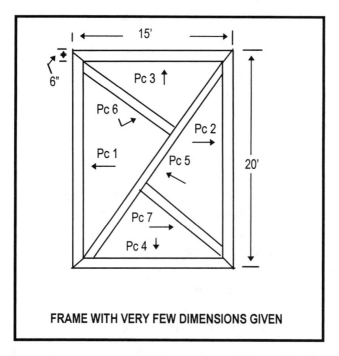

FRAME WITH VERY FEW DIMENSIONS GIVEN

An order came into a job shop where I was employed for two identical frames, which were 15′ by 20′. These frames were made of 3″×6″ rectangular tubing and were laid out as shown above. You will notice that there are not a lot of dimensions on this drawing. Since this was a rush order I was given one frame to build and another fabricator was given the other. We talked it over and I told him I would calculate the dimensions of each piece including those for cutting the mitered ends and would give them to him in less than an hour. He laughed and said there was no way to do that accurately enough to make it work. His solution was to sweep the floor in his work area and then he and a helper set to work making a full scale drawing on the floor with chalk from which he could measure the required dimensions and ultimately do the fit up.

Within an hour I had all the information I needed and started getting the materials together and cut. By the end of the day all the pieces were cut to length and mitered and I was ready to begin assembly. The other fabricator had finished his full scale drawing and was busy surrounding the area with horses and signs telling the night clean-up crew not to sweep away his full scale drawing. I find it much easier to do the calculations on a sheet of paper than on a 15′× 20′ floor. I will show you how to solve this problem completely after we have learned some facts about triangles.

ANGLES

The first thing we need to know is how angles are measured. If you look at a protractor you will see that a quarter circle is divided into 90 equal parts called degrees and that 90° represents two lines which are perpendicular to each other. If you are given an angle, you can measure it with a protractor. A word of caution, however. It's never good practice to try to get unknown measurements or angles by measuring off of the blueprints. You can make serious errors by assuming the drawing is accurate and to scale.

PARALLEL AND PERPENDICULAR LINES

Many projects involving steel have parallel lines in them. The edges of an angle iron, channel, wide flange or other beam are parallel and beams are often parallel to each other. When a line crosses two parallel lines it creates sets of angles which are equal to one another. Knowing which angles are equal can save a lot of time in calculating dimensions. The drawing below shows several sets of equal angles created by parallel lines. From here on I will write "∠ a" instead of "angle a."

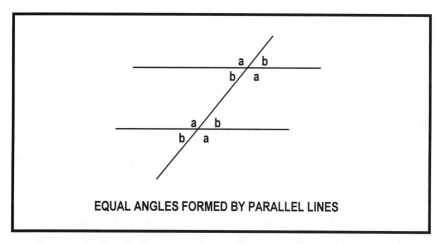

EQUAL ANGLES FORMED BY PARALLEL LINES

The drawing below shows an example using a beam. Note that since the beams A and B are perpendicular that the corner is a 90° angle so that $\angle$ b + $\angle$ a = 90° or 90° - $\angle$ a = $\angle$ b. This will be useful later on.

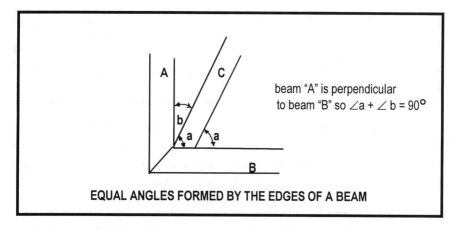

beam "A" is perpendicular to beam "B" so $\angle$a + $\angle$ b = 90°

EQUAL ANGLES FORMED BY THE EDGES OF A BEAM

Angles are often designated on a drawing by showing the "run" and the "rise" which represent that angle. The run shows how far to go horizontally from a given point and the rise shows how high to go vertically to create the desired angle. For example, a run of 12″ and a rise of 3″ is shown in the drawing below and represents an angle of about 14°.

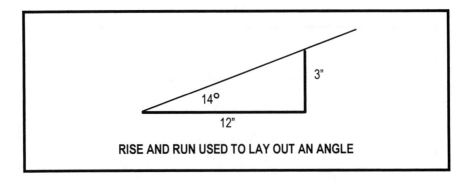

RISE AND RUN USED TO LAY OUT AN ANGLE

BASIC MATHEMATICS OF TRIANGLES

A right triangle is one in which two of the legs are perpendicular to each other, like the rise and the run. A right triangle is drawn as shown below.

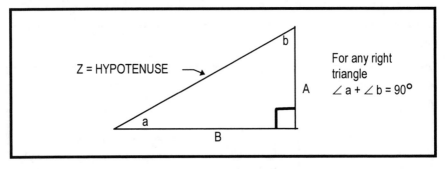

The box in the corner indicates the right (90°) angle and the side opposite this angle, Z, is called the hypotenuse. The other two angles will vary from triangle to triangle but for a right triangle there is an important connection between them.

For any triangle the sum of the angles is always 180°. For a right triangle, the sum of the degrees in the two non-right angles always equals 90°.

For example, if you know that one of the angles in a right triangle is 37° then the other angle is 90° - 37° = 53°.

Note that side A is opposite ∠ a and that side B is opposite ∠ b. Also note that side A is next to (adjacent to) ∠ b and that B is adjacent to ∠ a. When I talk about the sides of a right triangle I don't include the hypotenuse Z as one of the sides. This will be an important distinction in some of the formulas which follow.

TANGENTS AND MISSING DIMENSIONS

Referring back to the use of rise and run to designate an angle, note that a rise of 3″ in 12″ is the same as a rise of 6″ in 24″ or 9″ in 36″ and that they all represent the same angle. Also note that the two sides are always in the same proportion to one another. If we divide the rise by the run in each case we get $\frac{3}{12} = \frac{6}{24} = \frac{9}{36} = \frac{1}{4} = 0.25$. This means that the angle associated with this rise and run has the number 0.25 associated with it which is found by dividing the length of the side opposite the angle (the rise) by the length of the side adjacent to it (the run).

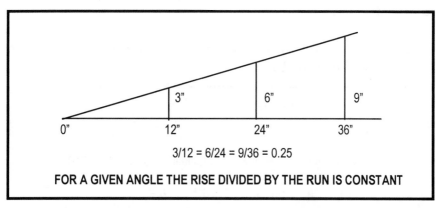

3/12 = 6/24 = 9/36 = 0.25

FOR A GIVEN ANGLE THE RISE DIVIDED BY THE RUN IS CONSTANT

To make it easier to visualize what we are doing we will not use the terms "rise" and "run" but will talk about the side opposite a given angle and the side adjacent to that angle. In this case if we divide the length of the side opposite the angle by the length of the side adjacent to the angle we always get a result of 0.25. **NOTE: All lenghts must have the same unit of measure before you divide them.** It is usually easiest to convert everything to inches.

It turns out that every angle has a unique number associated with it which is found by dividing the length of the side opposite the angle by the length of the side adjacent to the angle. Mathematically this ratio is called the "tangent" and is written as $\tan a = \frac{A}{B}$. It is abbreviated as TAN on a scientific calculator, which has tables of tangents built into it. If you know the angle you can enter that number in the calculator, press TAN and you will get a number that is the length of the side opposite divided by the length of the side adjacent. If you know the ratio of these sides, the tangent, you can enter it and than press **arctan**, (this key on your calculator may be labeled as ATAN or TAN^{-1}), and the number of degrees in the angle will appear. (CAUTION: Most scientific calculators allow you to measure angles in several different ways. Be sure you are in degree mode or your answers won't make any sense. You can check by finding **arctan** 1. If the answer is 45° you are in degree mode.) To remember what **arctan** means read it as "the angle whose tangent is." For example **arctan** (0.3456)= 19.06° can be read as "the angle whose tangent is 0.3456 is 19.06 degrees."

If you know the tangent and one of the sides you can calculate the missing dimension by using one of the following formulas.

The tangent of an angle, is found by dividing the length of the side opposite by the length of the side adjacent:

TANGENT = SIDE OPPOSITE divided by SIDE ADJACENT,

$$\tan a = \frac{A}{B} \text{ and } \tan b = \frac{B}{A}$$

You can then take that number and find the angle associated with it by using the **arctan** key.

ANGLE = arctan (SIDE OPPOSITE divided by SIDE ADJACENT)

Note: Divide the length of the side opposite by the length of the side adjacent and then use **arctan** with that number to get the angle

$$a = \arctan(A/B) \quad \text{and} \quad b = \arctan(B/A)$$

If you know the adjacent side and the angle in degrees you can find:

SIDE OPPOSITE = SIDE ADJACENT times TANGENT (a)

$$A = B(\tan a)$$

Note: B(tan a) means multiply the length of side B by the tangent of the ∠a.

If you know the opposite side and the angle in degrees you can find:

SIDE ADJACENT = SIDE OPPOSITE divided by TANGENT (a)

$$B = A/_{\tan a}$$

EXAMPLE: Given the triangle shown here, find ∠ a.

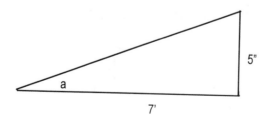

a

5"

7'

We want to know the angle, a, and we know that the side opposite the angle is 5″ and the side adjacent is 7′. (Note that the drawing is not to scale) If we say that $\tan a = \frac{5}{7} = .71429$ we will be making a serious error because the units of measure aren't the same. Convert 7′ to 84″ and we get $\tan a = \frac{5}{84}$. When you divide 5 by 84 on your scientific calculator you may get the strange answer of 5.95238...E-2. Don't let this scare you, the E-2 at the end means to move the decimal place two places to the left to give 0.05952, but you don't need to make this conversion because you only need this number to find the angle. Just press the arctan key to find arctan(5.95238...E-2). The result is about $3°$.

EXAMPLE: Given the triangle shown here, find the length of A.

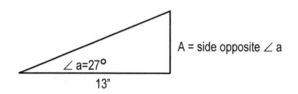

A = side opposite ∠ a

∠ a=27°

13″

We have the angle which is $27°$, and the side adjacent to it which is 13″. We use the formula: side opposite = side adjacent times tan (angle). Tan $27°$ = .50952. So side opposite = 13(.50952) = 6.6238. Remember that 13(.50952) means multiplication. Converting .6253 to 16ths we get the side opposite to be ~ $6\frac{10}{16}$ or $6\frac{5}{8}$″.

EXAMPLE: Given the drawing below, find the length of side B.

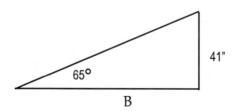

41″

65°

B

We have an angle of 65 degrees, the side opposite the angle is 41"so using side adjacent = side opposite divided by Tan (angle) we get side adjacent, B = $^{41}/_{tan\,65}$ = $^{41}/_{2.1147}$ = 19.1182 ≈ 19 $\frac{1}{8}$"

EXAMPLE: Given a rise of 3" for a run of 12" find ∠ a and ∠ b

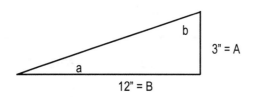

When we were looking at rise and run we had an example where the rise was 3" and the run was 12". We found that the tangent of the angle opposite the 3" long side was 0.25, so to find the angle whose tangent is 0.25 we use arctan 0.25 = 14.036° which for our purposes is 14°. We can find the other angle also. The side opposite ∠ b is 12 and the side adjacent is 3. $\tan b$ = $^{12}/_3$ = 4 and the angle whose tangent is 4, that is arctan 4 = 75.963°, which is 76° for our purposes. Notice that 14° + 76° = 90°. As I noted before the angles of a right triangle, not including the right angle, add up to 90 degrees.

USING SINES AND COSINES TO FIND MISSING DATA

There is also a useful relationship between the lengths of the side opposite the angle and the length of the hypotenuse (the side opposite the 90° angle). Referring to the triangle on page 56 the SINE of ∠ a is defined as the length of the side opposite ∠ a divided by the length of the hypotenuse Z. This is abbreviated as $\sin a$ = $^A/_Z$. Similarly $\sin b$ = $^B/_Z$. As with the tangent, if you know the value of the SINE you can use the arcsin (or ASIN, or SIN^{-1} depending on your calculator) to find the size of the angle.

60

Using the same reasoning as for the tangent we get the following formulas:

If you know the lengths of the hypotenuse and opposite side you can find:

SINE = SIDE OPPOSITE divided by HYPOTENUSE

$\sin a = {}^A/_Z$ **and** $\sin b = {}^B/_Z$

ANGLE = ASIN (SIDE OPPOSITE divided by HYPOTENUSE)

$a = \arcsin({}^A/_Z)$ **and** $b = \arcsin({}^B/_Z)$

If you know the length of the hypotenuse and the angle in degrees you can find:

SIDE OPPOSITE = HYPOTENUSE times SIN (angle)

$A = Z(\sin a)$

If you know the opposite side and the angle in degrees you can find:

HYPOTENUSE = SIDE OPPOSITE divided by SIN (angle)

$Z = {}^A/_{\sin a}$

Finally the COSINE of $\angle$ a, is defined as the length of the side adjacent to $\angle$ a divided by the length of the hypotenuse, Z. $\cos a = {}^B/_Z$ and $\cos b = {}^A/_Z$ This yields the following:

If you know the lengths of the hypotenuse and adjacent side you can find:

COS = SIDE ADJACENT divided by HYPOTENUSE

$\cos a = {}^B/_Z$ **and** $\cos b = {}^A/_Z$

ANGLE (degrees) = ACOS (SIDE ADJACENT divided by HYPOTENUSE)

$$a = \arccos\left(\frac{B}{Z}\right) \text{ and } b = \arccos\left(\frac{A}{Z}\right)$$

If you know the hypotenuse and the angle in degrees you can find:

SIDE ADJACENT = HYPOTENUSE times COS (angle)

$$B = Z\cos a$$

If you know the adjacent side and the angle in degrees you can find:

HYPOTENUSE = SIDE ADJACENT divided by COS (angle)

$$Z = \frac{B}{\cos a}$$

EXAMPLE: Given the triangle shown below with an angle of 20 degrees and a rise of 5 feet find the run B.

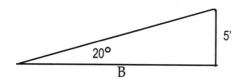

Since we know $\angle$ a and the length of the side opposite and want to find the side adjacent we will use the tangent relationship $B = \frac{A}{\tan a}$. We know the length of side A opposite the angle is 5' and we know the $\angle$ a=20°. First convert 5' to 60", then use your calculator to find tan 20 = .36397.

$$B = \frac{A}{\tan a} = \frac{60}{.36397} = 164.8486$$

We will convert this to feet and inches by dividing by 12 to get 13.7374'. Multiply .7374 by 12 to get 8.8488". Multiply .8488 by 16 to get 14 which means we have $^{14}\!/_{16}$ or $\frac{7}{8}$. So the result is ~13'–8$\frac{7}{8}$".

EXAMPLE Given the triangle shown below we want to find the rise and the run.

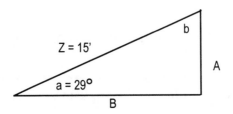

We know a = 29° Z = 15'. We want to find A and B. To find A we use the sine relationship:

$$A = Z(\sin a) = 15(\sin 29) = 15(.48481) = 7.2721$$

Since this is in feet we convert the decimal part to inches and 16ths. .2721(12) = 3.2657 " and .2657(16) ≈ 4.25 so the result is $A \approx 7'–3\frac{4}{16}$" or $7'–3\frac{1}{4}$".

Similarly
$$B = Z(\cos a) = 15(\cos 29) = 15(.87462) = 13.1193$$

Converting to inches and 16ths we get ~13'–1$\frac{1}{16}$".

PYTHAGOREAN THEOREM

There is another relationship between the lengths of the sides of a right triangle which we can use in fabrication, but before we go on we need a couple of definitions. If we multiply 5, by itself we say that we have squared the number 5 because the result is the area of a square whose sides are 5 units long. Mathematically we write this as $5^2 = 25$. Similarly if we have a number and want to know what number multiplied by <u>itself</u> will give that number we are looking for the square root. For example: 4 squared is the same as 4×4 and is written as 4^2 =16. The square root of 16 is 4 since 4×4=16 and is written as $\sqrt{16}$ = 4 . Note that there are other combinations of numbers that can be multiplied together to give 16. For example 8×2=16 but neither 8 nor 2 is the square root of 16 since neither 2×2 nor 8×8 equals 16. What if we want the square root of 13? We know that 3×3=9 and 4×4=16 so the square root must be somewhere in between. Most numbers do not have whole number square roots so we use the scientific calculator. Enter the number 13 and then press the square root key and you get 3.606. Sure enough 3.606×3.606=13.003 which is the same as 13 if we round off the numbers.

Pythagorus discovered that, for any right triangle, if you make a square out of each of the sides, the area of the square created by the hypotenuse equals the sum of the areas of the squares created by the legs(see the picture below).

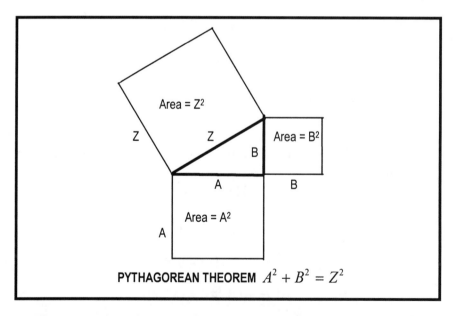

This says that if we multiply the length of A by itself and the length of B by itself and add the two together we get the same number as when we multiply the length of Z by itself. Using this formula we can derive several other formulas which are also very useful.

$Z = \sqrt{A^2 + B^2}$ Given the lengths of sides A and B, multiply each by itself, add them together and then take the square root. The result is the length of the hypotenuse.

$A = \sqrt{Z^2 - B^2}$ Given the lengths of the hypotenuse and side B, square each and subtract . Take the square root of the result to get the length of side A.

$B = \sqrt{Z^2 - A^2}$ Given the length of the hypotenuse and side A, square each and subtract. Take the square root to get the length of side B.

EXAMPLE: Given the triangle below find the length of Z.

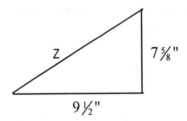

First convert the fractions to decimals by dividing the numerator by the denominator.

$$\tfrac{5}{8} = .625 \text{ and } \tfrac{1}{2} = 0.5$$

$$Z = \sqrt{(9.5)^2 + (7.625)^2}$$

$$(9.5)^2 = 90.25;\ (7.625)^2 = 58.141$$

$$90.25 + 58.141 = 148.391$$

$$\sqrt{148.391} = 12.1816 \approx 12\tfrac{3}{16}$$

EXAMPLE: Given the triangle below find the length of A.

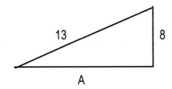

$A = \sqrt{13^2 - 8^2}$; 13^2=169; 8^2=64; 169 - 64=105; $\sqrt{105}$ = $10\frac{1}{4}$.

There are some useful combinations where all the sides are whole numbers. Two useful ones are:

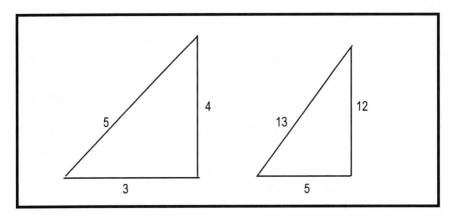

These triangles can be used to make sure one line is perpendicular to another in a large lay out. If you need a larger triangle you can double or triple each side of the 3-4-5 triangle to get a 6-8-10 or a 9-12-15 triangle or, if you need a smaller triangle you can divide each leg of a 3-4-5 by 2 to get a 1.5-2-2.5 triangle which is still a right triangle. (Use your calculator to show that 1.5^2 + 2^2 = 2.5^2.) For example, if a piece that is 14′ long has to be perpendicular to a piece 17′ long you use a multiple of a 3-4-5 triangle, for example a 12-16-20. This will be much more accurate than using a framing square.

Say we want to build a structure like that shown below.

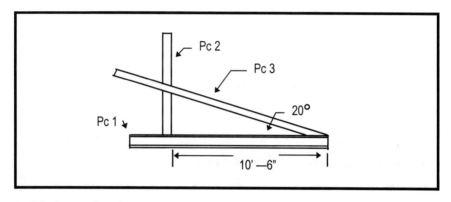

We have the three pieces and need to make sure that piece 2 is perpendicular to piece 1 and that piece 3 is set at 20°. First set piece 2 at a distance of 10′−6″ from the end of piece 1 as shown. The location of this piece is fixed because it is specified on the drawing. We know that a 3-4-5 triangle (and all its multiples) is a right triangle and we will use this fact to make sure piece 2 and piece 1 are perpendicular. We can use a 6-8-10 for this structure so mark a point 8′ from the fixed point on piece 1 and a point 6′ from the fixed point on piece 2. Adjust the top of piece 2 (don't move the bottom off the fixed point) until the distance between the 8′ mark and the 6′ mark is exactly 10′. Piece 2 is now perpendicular to piece 1.

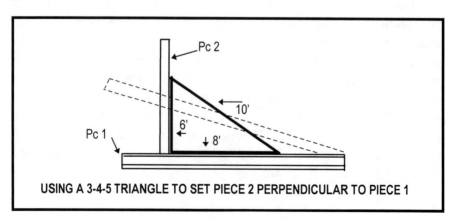

USING A 3-4-5 TRIANGLE TO SET PIECE 2 PERPENDICULAR TO PIECE 1

We now want to place piece 3, the inclined beam which welds to the back of the vertical beam. We know from the picture that the toe of the inclined beam is to be $10' - 6''$ from the upright beam and it is inclined at 20°. One way you could do this is to use a magnetic angle finder as described in the tool section. This would be fine if the location of the beam is not critical, but remember that the magnetic angle finder sits on a very small base and if the beam is bowed the reading will be different in different places.

A more accurate way is to figure out how high up on the vertical beam the inclined beam should cross and set the beam to that height. Although the distance from the toe of the inclined beam to the point where it crosses the vertical beam is not essential we will calculate it anyway and use it as a cross check when we set the beam.

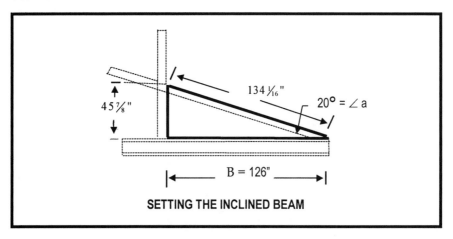

SETTING THE INCLINED BEAM

It's usually easier to work in inches when calculating so lets convert $10' - 6''$ to $126''$. From the formulas we see that we have $\angle a$ and side B and that side A can be found using $A = B(\tan a)$ so $A = 126(\tan 20)$. Use your calculator to find the tangent of 20 degrees which is 0.36397. Now multiply this times 126 to get 45.86″. To get this in inches and 16ths multiply the decimal part, .86 times 16 and get 13.76 which is close to $^{14}\!/_{16}$ or $\frac{7}{8}$. Note that $45\frac{7}{8}''$ is where the top edge of the beam crosses the vertical beam.

To find the distance from the toe of the inclined beam to the point where it crosses the vertical beam use the formula $z = \sqrt{A^2 + B^2}$. We are given B which is 126". We just found A which is 45.86". A^2=2103.14. B^2=15876 so A^2+B^2=17979.14. Now we take the square root of this number using our calculator again and get 134.09. .09 times 16 equals 1.44 or about $\frac{1}{16}$. So the distance from the toe of the inclined beam to the point where it crosses the vertical beam is ~134$\frac{1}{16}$". Again this is measured along the top surface.

Congratulations! If you have worked through the examples given and understand the concepts you now have a basic understanding of trigonometry. We will move on to figuring out the lengths of all the pieces in the frame and the dimensions of the miter cuts so all the pieces will fit.

APPLIED MATH
SOLVING A PRACTICAL PROBLEM

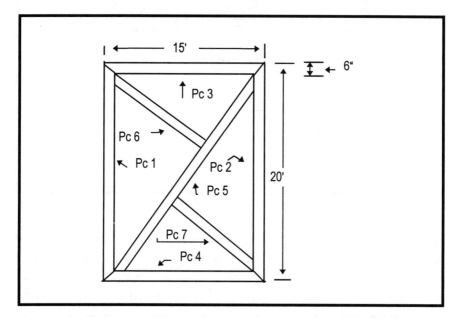

We will now go back to our original problem and use our understanding of triangles to get the information we need. From the drawing we see that we really have very little information. We know the frame is 15' by 20' and that the tubing is 6" wide. The lengths of some of the pieces are fairly obvious but for other pieces we have no idea how long they are. The same is true of the angles. The problem will be to calculate the lengths of material required and then to calculate the angles and how to lay the miter cuts out on each piece.

Pieces 1, 2, 3, and 4 are fairly simple. Pieces 1 and 2 are 20′ long and pieces 3 and 4 are 15′ long. Since the corners of the frame are right angles (90°) the ends of these pieces are mitered at 45°. This miter can either be laid out using a combination square or the cuts can be laid out by measurement. For a 45° angle both legs are equal. Since the width of the tubing is 6″ we can measure back 6″ as shown below and go from there.

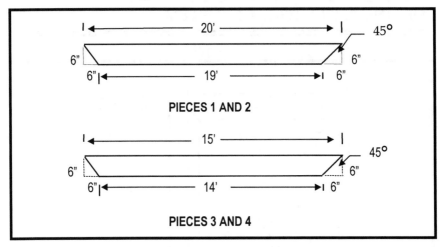

Before we go on lets look at the angles to see if there is any pattern that we can use to our advantage. Look at the triangle at the bottom of the drawing which contains the 14′ inside edge of piece 4, piece 7 and part of piece 5. Piece 7 meets piece 5 at 90°. Call the other two angles ∠ a and ∠ b and note that ∠ a+∠ b = 90°.

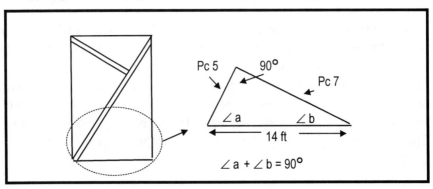

Now, looking at the drawing below we see that piece 5 is a pair of parallel lines crossed by a diagonal line (piece 4). This means that the two angles formed by pieces 4 and 5 are equal. The lower left corner of the frame is a right angle and is the sum of ∠ a and the angle formed by the 19′ long inside edge of piece 1 and piece 5. But we know that ∠ a + ∠ b = 90° so that lower angle must be ∠ b.

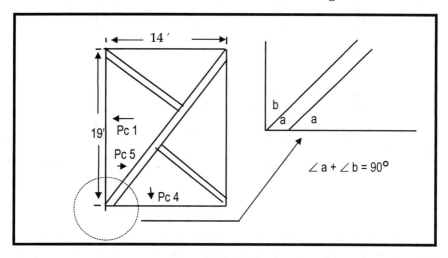

Now looking at the triangle made up of the 19′ inside edge of piece 1, piece 6 and part of piece 5 we see that the upper angle must be ∠ a, since it plus ∠ b equals 90°. By following the angles around the frame we see that the angles are as marked below.

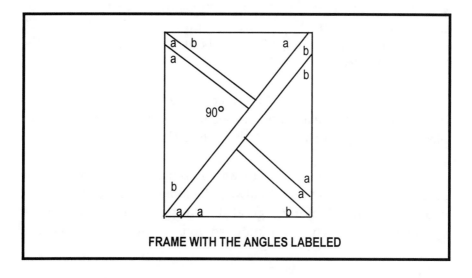

FRAME WITH THE ANGLES LABELED

Since we only have the two angles $\angle$ a and $\angle$ b lets figure them out so we can use them. Look at the large triangle made of the 14′ inside edge of piece 4, the 19′ inside edge of piece 2 and piece 5.

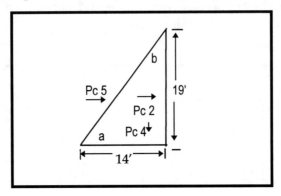

we see that $\tan a = \frac{19}{14} = 1.35714$ and the angle is $a = \arctan 1.35714 = 53.62$. Since $\angle$ a+$\angle$ b = 90° we have that $\angle$ b = 90 - 53.62 = 36.38°.

From the above drawing we see that piece 5 is the hypotenuse of a right triangle with a 14′ side (the inside of piece 4) and a 19′ side (the inside edge of piece 2). Since we know two sides of a right triangle we can use the Pythagorean theorem.

Using the formula $z = \sqrt{A^2 + B^2}$ we get $z = \sqrt{19^2 + 14^2}$. Using your calculator 19²=361; 14²=196 and 19²+14²=557. So $z = \sqrt{557}$ which is 23.6′. Multiplying .6 by 12 we get 7.2″. Multiplying .2 by 16 we get 3.36 or about $\frac{3}{16}$. So the length of the long side of piece 5 is ~ 23′–7 $\frac{3}{16}$″.

We could also have used the fact that $Z = \frac{19}{\sin 53.6}$ to check this dimension. Do this calculation to see if you get the same length for piece 5 that we got using the Pythagorean Theorem.

A quick note on procedure and notation: when doing these kinds of calculations you will always be looking for something that you don't know. Making a good drawing is important because it will help you see what you know so you can figure out what you don't know. Label the drawing with the information you <u>do</u> know and label the unknown items with some code that you can come back to later and understand. Notice that in my drawings I have shown lengths between points clearly and that I have used capital letters for lengths of sides and small letters for angles. This is comfortable to me because I can come back to a drawing a week later and remember what I did. In the calculations below I have labeled the lengths I am looking for with the letters A, B, and Z and the angles as a and b. I will solve the big problem as a series of individual smaller problems.

Look at the following drawing to see how the miters are calculated for piece 5.

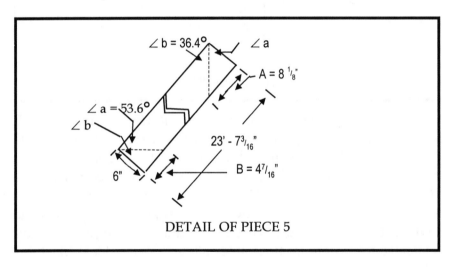

DETAIL OF PIECE 5

The ends of the tubing to make piece 5 are right angles before they are mitered and we know that the angle of the upper cut is ∠ b = 36.32°. We want the length, A, from the end of the beam to the end of the miter so we look at the triangle which will be cut off. For that triangle we have the ∠ a = 53.62° and the side adjacent is the width of the tubing which is 6". We use $A = 6(\tan 53.62) = 8.14415$" or approximately $8\frac{1}{8}$".

For the lower cut we have the $\angle$ b = 36.38° and the side adjacent is 6 inches. We use $B = 6(\tan 36.38)$ which is 4.420 or approximately $4\frac{7}{16}''$.

Now for piece 6. Look at the drawing below.

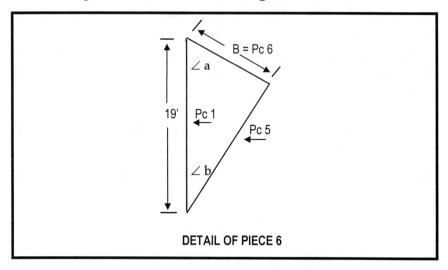

DETAIL OF PIECE 6

To calculate the length of piece 6 we will need to use the SINE function because the only length we have is the side opposite the 90 degree angle. Using $\angle$ b in the lower corner, the hypotenuse which is 19 feet and calling the length we need to know as B, we have $B = 19\sin(36.4)$ = 11.275'. Converting to feet and inches we get that piece 6 is ~$11'$-$3\frac{5}{16}''$. Now for the angle cut. Note that the angle in the corner of the piece is $\angle$ a and that we have already calculated the lay out for this cut for piece 5. Piece 6 looks like this:

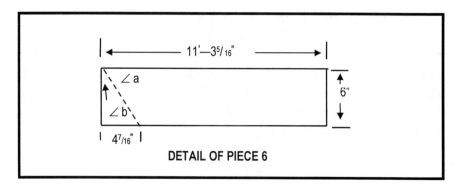

DETAIL OF PIECE 6

Piece 7 is calculated the same way except that the lengths are different because the diagonal brace, piece 5, is offset. Use the figure below and refer to the figure for piece 5.

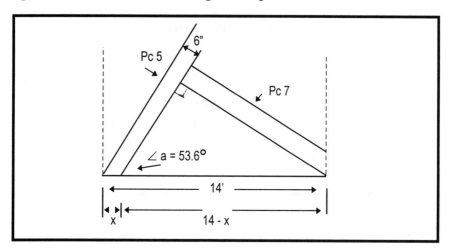

The problem is that we don't have a complete right triangle, but there are two ways to create one. The easiest way is to extend piece 7 across piece 5 to create a triangle with the hypotenuse of 14 feet and the angles which we already know. The other way is to subtract the length, x, of the diagonal of the miter cut of piece 5 from the 14' length and then use the angles. I will do it both ways as a cross-check of the result.

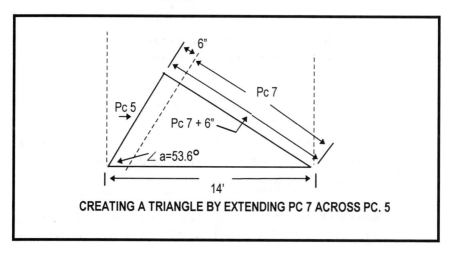

CREATING A TRIANGLE BY EXTENDING PC 7 ACROSS PC. 5

We know that the tubing is 6″ wide so our triangle has one side which is equal to piece 7 plus 6″. We will have to remember to subtract this extra length to get piece 7. Since 14′ = 168″ we have piece 7 +6″ = 168 sin 53.6. If we subtract 6 from each side we get

$$\text{piece } 7 = (168 \sin 53.6) - 6'' = 129.222'' = 10'-9\tfrac{1}{4}''$$

To use the other method we need to calculate the length of the miter cut on piece 5 so that we can subtract it from 14 feet to get the length, Z, of the hypotenuse of the inside triangle. We know the two sides of the miter — they are $4\tfrac{7}{16}''$ and 6″.

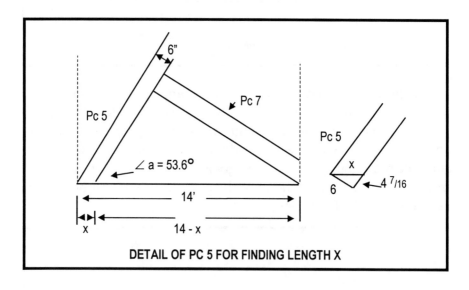

DETAIL OF PC 5 FOR FINDING LENGTH X

At this point it is wise to remember that the $4\tfrac{7}{16}$ dimension was derived from a decimal dimension which was calculated. This original dimension was 4.421 while $4\tfrac{7}{16} = 4.43754$. It is safer to use the original number so that you don't accumulate error.

$$x = \sqrt{6^2 + 4.420^2} = \sqrt{55.536} = 7.4522 .$$

Now we need to be careful not to mix up feet and inches. 14′ is 168″. Length $Z = 168 - 7.4529 = 160.5477''$. From the picture below $A = Z \sin a = 160.54773(\sin 53.62) = 129.257$ which is $\sim 129\,\frac{1}{4}''$ or $10'–9\,\frac{1}{4}''$, which agrees with the other calculation.

Finally referring to the drawing again we see the cutting angle is $\angle$ a. Using the calculations for piece 5 we see that piece 7 looks like

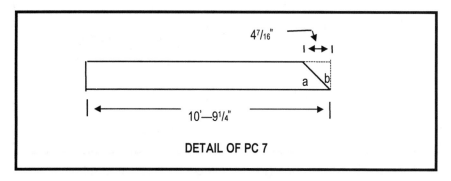

DETAIL OF PC 7

We have now calculated all the vital information we need to cut the tubing to length and cut the angles. It took me about an hour to make the drawings, do the calculations and double check the results. If it took three times that long it would be better than trying to draw the whole thing out on the floor. For one thing you can do it alone whereas doing a floor lay out of this size would take two people to make sure the drawing was square and accurate. Your calculated results will probably be more accurate than those measured from a full or part scale drawing. Finally, the more you can figure out on your own the more valuable you are to any shop you work in and the more responsibility you will get as time goes by.

Now that the pieces are all cut to length the next steps involve assembling the frame. Using the techniques taught earlier in this book you will be able to make sure the frame is both level and square.

CIRCLES AND FLANGE LAY-OUT

Circles are not as commonly used as triangles in metal fabrication but the basic relationships between the parts is useful.

The basic formulas are:

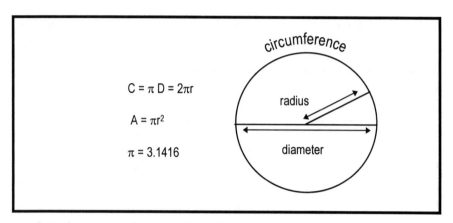

$$C = \pi D = 2\pi r$$

$$A = \pi r^2$$

$$\pi = 3.1416$$

C is the circumference (the distance around the circle), A is the area, D is the diameter (the distance across the circle through the center), r is the radius (half of the diameter) and π (pi) is equal to 3.14159. If you need to make an 8″ diameter circle out of flat bar you will need at least $C = \pi \times 8 = 8 \times 3.14159 = 25.113″$ of material.

You may be called upon to find the center of an existing circle. This comes up in laying out pipe flanges or hubs of different types. If you cut the circle out yourself you will have drawn it around a known center, but if you are given a circle that was cut elsewhere you will have to find the center before you can begin. There is a geometric fact which will help you here. Draw a line across the circle so that it is a reasonable distance away from the center. Find the center point of this line using your tape measure or a compass.[*] Draw a line perpendicular to the line at this point. This second line will pass through the center of the circle. If we follow this procedure again using a line which is not parallel to the first line we will get another perpendicular. Where the two perpendiculars cross is the center of the circle. To double check yourself always draw a third line and its perpendicular to make sure it goes through the same point.

---

[*] See appendix for information on doing geometric constructions.

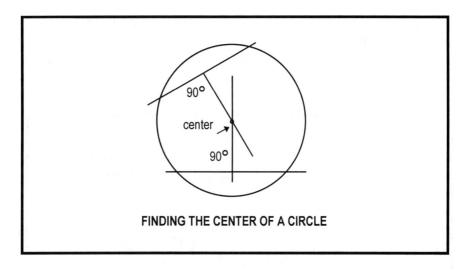

FINDING THE CENTER OF A CIRCLE

Finally you will need to know that there are 360° in a circle. This can be seen by adding up the 90 ° angles in the figure below.

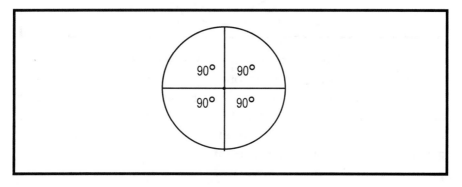

We will now use our knowledge of triangles and circles to lay out the bolt hole circle on a flange. We are given the blueprint for two flanges as shown on the next page together with two pieces of steel which are the correct diameter and with the center holes already in them. Each flanges will be welded to a pipe and then the pipes will be joined together by bolts through the flanges. The best way to make sure the holes will line up is to tack them together, lay out the holes and drill them both at one time.

First we need to draw the bolt circle which requires finding the center of the circle. Find a piece of scrap to tack over the center hole and then use the method shown earlier to find the center of the circle. Center punch that location and use a compass to draw a 16″ diameter (8″ radius) circle. My preference is to use a compass with a carbide tip. This will scratch a finer line than a chalk or silver pencil and is more accurate.

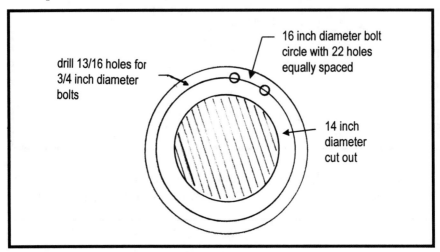

Next we need to figure out how to locate the holes by finding the straight line distance from one hole to the next. To calculate this distance we will first draw a picture, see below. The diameter of the bolt circle is 16 inches so the radius is 8 inches. Also since there are 360° in a circle and we are dividing it into 22 sectors we get 360 ÷ 22 = 16.3636° in each sector.

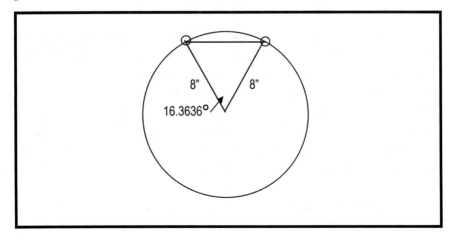

Note that the triangle we have constructed is not a right triangle but that we can construct a right triangle by drawing a line from the center of the circle to the mid-point of the line between the holes. This new line also divides the 16.3636° angle in half creating two angles of 8.1818°.

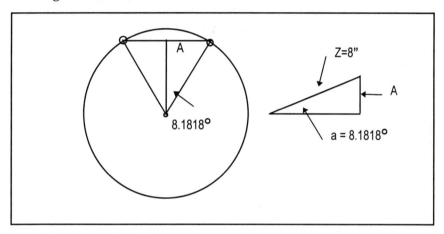

We will call the length of the side opposite the 8.1818° angle by the letter A and remember that it is half the distance between the holes. From our work with triangles we have that $A = 8\sin 8.1818 = 8(.14231) = 1.1385$. Since this is half the distance we get that the straight line distance between the holes is about 2.277 ". Converting .227 to 32nds we get 8.86 or approximately $^9/_{32}$, so the straight line distance between the hole centers is a little less than $2\,^9/_{32}$"

To finish the lay-out make a punch mark on the bolt circle to locate the first hole. CAUTION: Don't center punch any hole but the first one until you are all finished.) It doesn't make any difference where on the circle you locate the first hole. Now open your compass to "and mark the second hole and use that mark to locate the third hole and so on all the way around the circle. Remember! Don't do any center punching yet! Now check to see if the last hole marked is $2\,^9/_{32}$" from the center punch marking the first hole. If it is too short or too long adjust your compass accordingly and re-mark the holes until they all come out equally spaced. There will likely be an error caused by rounding 8.86 up to 9 when we were finding the fractional equivalent of .227. When you've got the lay-out correct, center punch for drilling. Use a permanent marker to draw a small circle around each center punch mark so they are easy to find and write the hole size on the face of the flange. Finally use your punch to mark the edge of the flanges so that they can be realigned after they are separated. I put the punch marks off center so that the matching faces can be easily seen as shown below.

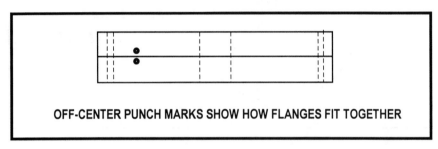

OFF-CENTER PUNCH MARKS SHOW HOW FLANGES FIT TOGETHER

If you are going to drill them I recommend using a quarter inch drill as a pilot to make sure that the drilled holes end up where they are supposed to be.

So now you have the basics. As you work in the shop you will learn other techniques which will help you solve problems that arise on a daily basis. The main tools you have for solving problems are your brain, eyes and ears. Don't be afraid to learn new things. Mathematics is a tool just like any other. Trust your eyes and look over your work to make sure it's right. Listen to those with experience. They have a wealth of information which is probably not written down anywhere. You'll never be wasting your time by learning as much as you can.

APPENDIX

DIMENSIONS OF COMMON STEEL SHAPES

CHANNEL flange width 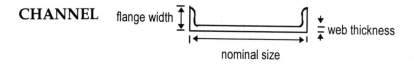 web thickness

nominal size

SIZE inches	WEIGHT pounds per foot	FLANGE WIDTH inches	WEB THICKNESS inches
3	4.1	1.41	.170
	5.0	1.50	.258
	6.0	1.60	.356
4	5.4	1.58	.180
	6.25	1.65	.247
	7.25	1.72	.320
5	6.7	1.75	.190
	9.0	1.89	.325
6	8.2	1.92	.200
	10.5	2.03	.314
	13.0	2.16	.437
7	9.8	2.09	.210
	12.25	2.19	.314
	14.75	2.30	.419
8	11.5	2.26	.220
	13.75	2.34	.303
	18.75	2.53	.487
9	13.4	2.43	.230
	15.0	2.49	.285
	20.0	2.65	.448
10	15.3	2.60	.240
	20.0	2.74	.379
	25.0	2.89	.526
	30.0	3.03	.673
12	20.7	2.94	.280
	25.0	3.05	.387
	30.0	3.17	.510
15	33.9	3.40	.400
	40.0	3.52	.520
	50.0	3.72	.716

SIZE inches	WEIGHT pounds per foot	FLANGE WIDTH inches	WEB THICKNESS inches
MC6	12.0	2 -1/2	5/16
	15.1	3	5/16
	15.3	3 -1/2	5/16
	16.3	3	3/8
	18.0	3 -1/2	3/8
MC7	19.1	3-1/2	3/8
	22.7	3-5/8	1/2
MC8	18.7	3	3/8
	20	3	3/8
	21.4	3-1/2	3/8
	22.8	3-1/2	7/16
MC9	23.9	3-1/2	3/8
	25.4	3-1/2	7/16
MC10	21.9	3-1/2	5/16
	24.9	3-3/8	3/8
	25.3	3-1/2	7/16
	28.3	3-1/2	1/2
	28.5	4	7/16
	33.6	4-1/8	9/16
	41.1	3-3/8	13/16
MC12	30.9	3-1/2	7/16
	32.9	3-1/2	1/2
	35	3-3/4	7/16
	37	3-5/8	5/8
	40	3-7/8	9/16
	45	4	11/16
	50	4-1/8	13/16
MC13	31.8	4	3/8
	35	4-1/8	7/16
	40	4-1/8	9/16
	50	4-3/8	13/16

ANGLE IRON

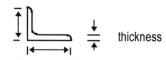 thickness

SIZE inches	THICK NESS inches	WEIGHT pounds per foot
1/2x1/2	1/8	0.38
3/4x3/4	1/8	0.59
1x1	1/8	0.80
	3/16	1.16
	1/4	1.49
1-1/4x1-1/4	1/8	1.01
	3/16	1.48
	1/4	1.92
1-1/2x1-1/4	3/16	1.64
1-1/2x1-1/2	1/8	1.23
	3/16	1.80
	1/4	2.34
	5/16	2.86
1-3/4x1-3/4	1/8	1.44
	3/16	2.12
	1/4	2.77
2x1-1/2	1/8	1.44
	3/16	2.12
	1/4	2.77
2x2	1/8	1.65
	3/16	2.44
	1/4	3.19
	5/16	3.92
	3/8	4.70
2-1/2x1-1/2	3/16	2.44
	1/4	3.19
2-1/2x2	3/16	2.75
	1/4	3.62
	5/16	4.50
	3/8	5.30
2-1/2x2-1/2	3/16	3.07
	1/4	4.10
	5/16	5.00
	3/8	6.90
	1/2	7.70
3x2	3/16	3.07
	1/4	4.1
	5/16	5.0
	3/8	5.9
	1/2	7.7
3x2-1/2	1/4	4.5
	5/16	5.6
	3/8	6.6
	1/2	8.5
3x3	3/16	3.71
	1/4	4.9
	5/16	6.1
	3/8	7.2
	7/16	8.3
	1/2	9.4
3-1/2 x 2-1/2	1/4	4.9
	5/16	6.1
	3/8	7.2
	1/2	9.4

SIZE inches	THICK NESS inches	WEIGHT pounds per foot
3-1/2 x 3	1/4	5.4
	5/16	6.6
	3/8	7.9
	1/2	10.2
3-1/2 x 3-1/2	1/4	5.8
	5/16	7.2
	3/8	8.5
	7/16	9.8
	1/2	11.1
4x3	1/4	5.8
	5/16	7.2
	3/8	8.5
	7/16	9.8
	1/2	11.1
	5/8	13.6
4x3-1/2	1/4	6.2
	5/16	7.7
	3/8	9.1
	1/16	10.6
	1/2	11.9
4x4	1/4	6.6
	5/16	8.2
	3/8	9.8
	7/16	11.3
	1/2	12.8
	5/8	15.7
	3/4	18.5
5x3	1/4	6.6
	5/16	8.2
	3/8	9.8
	7/16	11.3
	1/2	12.8
5x3-1/2	1/4	7.0
	5/16	8.7
	3/8	10.4
	7/16	12.0
	1/2	13.6
	5/8	16.8
	3/4	19.8
5x5	5/16	10.3
	3/8	12.3
	7/16	14.3
	1/2	16.2
	5/8	20.0
	3/4	23.6
6x3-1/2	5/16	9.8
	3/8	11.7
	1/2	15.3
6x4	5/16	10.3
	3/8	12.3
	7/16	14.3
	1/2	16.2
	5/8	20.0
	3/4	23.6
	7/8	27.2

WIDE FLANGE

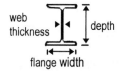

web thickness | depth | flange width

SIZE inches	WEIGHT pounds per foot	DEPTH inches	FLANGE WIDTH inches	WEB THICK NESS inches	SIZE inches	WEIGHT pounds per foot	DEPTH inches	FLANGE WIDTH inches	WEB THICK NESS inches
4	13	4.16	4.060	.280	14	22	13.74	5.00	.230
5	16	5.01	5.00	.240		26	13.91	5.025	.255
	19	5.03	5.27	.240		30	13.84	6.73	.270
6	9	5.90	3.94	.170		34	13.98	6.745	.285
	12	6.03	4.00	.230		38	14.10	6.77	.310
	15	5.99	5.99	.230		43	13.66	7.995	.305
	16	6.28	4.03	.260		48	13.79	8.03	.340
	20	6.2	6.02	.260		53	13.92	8.06	.370
	25	6.38	6.08	.320		61	13.89	9.995	.375
8	10	7.89	3.94	.170		68	14.04	10.035	.415
	13	7.99	4.00	.230		74	14.17	10.07	.450
	15	8.11	4.015	.245		82	14.31	10.13	.510
	18	8.14	5.25	.230		90	14.02	14.52	.440
	21	8.28	5.27	.250		99	14.16	14.565	.485
	24	7.93	6.495	.245		109	14.32	14.605	.525
	28	8.06	6.535	.285		120	14.48	14.67	.590
	31	8.00	7.995	.285		132	14.66	14.725	.645
	35	8.12	8.02	.310		145	14.78	15.50	.680
	40	8.25	8.07	.360		159	14.98	15.565	.745
	48	8.50	8.11	.400		176	15.22	15.65	.830
	58	8.75	8.22	.510	16	26	15.69	5.50	.250
	67	9.00	8.28	.570		31	15.88	5.525	.275
10	12	9.87	3.96	.190		36	15.86	6.985	.295
	15	9.99	4.00	.230		40	16.01	6.995	.305
	17	10.11	4.01	.240		45	16.13	7.035	.345
	19	10.24	4.02	.250		50	16.26	7.07	.380
	22	10.17	5.75	.240		57	16.43	7.12	.430
	26	10.33	5.77	.260		67	16.33	10.235	.395
	30	10.47	5.81	.300		77	16.52	10.295	.455
	33	9.73	7.96	.290		89	16.75	10.365	.525
	39	9.92	7.985	.315		100	16.97	10.425	.585
	45	10.10	8.02	.350	18	35	17.70	6.00	.300
	49	9.98	10.00	.340		40	17.90	6.015	.315
	54	10.09	10.03	.370		46	18.06	6.06	.360
	60	10.22	10.08	.420		50	17.99	7.495	.355
	68	10.40	10.13	.470		55	18.11	7.53	.390
	77	10.60	10.19	.530		60	18.24	7.555	.415
	88	10.84	10.265	.605		65	18.35	7.59	.450
	100	11.10	10.34	.680		71	18.47	7.635	.495
	112	11.36	10.415	.755		76	18.21	11.035	.425
12	65	12.12	12.00	.390		86	18.39	11.09	.480
	72	12.25	12.04	.430		97	18.59	11.145	.535
	79	12.38	12.08	.470		106	18.73	11.20	.590
	87	12.53	12.125	.515		119	18.97	11.265	.655
	96	12.71	12.16	.550	21	44	20.66	6.50	.350
	106	12.89	12.22	.610		50	20.83	6.53	.380
	120	13.12	12.32	.710		57	21.06	6.555	.405
	136	13.41	12.40	.790		62	20.99	8.24	.400
	152	13.71	12.48	.870		68	21.13	8.27	.430
	170	14.03	12.57	.960		73	21.24	8.295	.455
	190	14.38	12.67	1.06		83	21.43	8.355	.515

GEOMETRIC CONSTRUCTIONS

CONSTRUCTING THE PERPENDICULAR BISECTOR OF A LINE

The perpendicular bisector of a line consists of all the points which are equidistant from the end points. To find these points you do the following.

1. Place the point of the compass on one end of the line and open it until it reaches beyond the center of the line.
2. Draw arcs above and below the line using the end points as centers The points at which the arcs cross are the same distance from either end.
3. Use a straight edge to draw a line between these points of intersection to create a line which is perpendicular to the given line and bisects it.

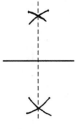

If you are given a point on a line and want to construct a line that goes through that point and is perpendicular to the line, open your compass and make arcs which cross the given line. These new points are like the end points in the method above so go back to the previous instructions

BISECTING AN ANGLE.

Given the angle, put the point of your compass at the vertex and make an arc that crosses both legs. Now put the point of the compass at each of these points of intersection and make two new arcs which intersect. Draw a line through the vertex and the intersection of the arcs to create the bisector of the angle.

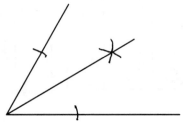

FORMULAS

PYTHAGOREAN THEOREM:

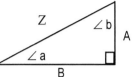

$$A^2 + B^2 = Z^2$$

For any triangle the sum of the angles
is 180°. **For a right triangle, $\angle a + \angle b = 90°$.**

TANGENT OF AN ANGLE

Tangent ($\angle$ a) = length A divided by length B   **tan (a) = A / B.**
                                                      tan (b) = B / A

Angle "a" is the angle whose tangent is A / B   **$\angle$ a = arctan (A / B)**
                                                      $\angle$ b = arctan (B / A)

SINE OF AN ANGLE

Sine ($\angle$ a) = length A divided by length Z   **sin (a) = A / Z**

                                                      sin (b) = B / Z

Angle "a" is the angle whose sine is A / Z   **$\angle$ a = arcsin (A / Z)**
                                                      $\angle$ b = arcsin (B / Z)

COSINE OF AN ANGLE

Cosine ($\angle$ a) = length B divided by length Z   **cos (a) = B / Z**

                                                      cos (b) = A / Z

Angle "a" is the angle whose cosine is A / B   **$\angle$ a = arccos (A / Z)**
                                                      $\angle$ b = arccos (B / Z)

CIRCLES

Circumference $C = \pi D = 2\pi r$
Area $A = \pi r^2$,
$\pi = 3.14159$

INDEX

HOW TO BUILD WITH STEEL

A book written for full time steel fabricators, students, or anyone with a week-end project

- **Tools, including techniques used in steel fabrication**

- **Practical procedures for lay-out, fit-up, and material preparation**

- **Use of the torch for cutting material and removing welding distortion**

- **The mathematics of angles, triangles, and circles and applications to building steel structures**

- **Tables of specifications for structural steel**

ORDER FORM

If you can't find a copy at your local bookstore, additional copies of *HOW TO BUILD WITH STEEL* are available from Shire Press for ($10) per copy plus $1.50 per copy for postage and handling.

Please send_____ copies of *HOW TO BUILD WITH STEEL* to

_____

_____

_____

Please enclose payment with your order. California residents please add sales tax.

SHIRE PRESS 26873 Hester Creek Road, Los Gatos, CA 95033